AF522272

ENCYCLOPAEDIC DICTIONARY OF
CHEMISTRY

ENCYCLOPAEDIC DICTIONARY OF CHEMISTRY

Vol. 2
Physical Chemistry

Edited by
SATISH ANAND
RAJ KUMAR

ANMOL PUBLICATIONS PVT. LTD.
NEW DELHI - 110 002 (INDIA)

ANMOL PUBLICATIONS PVT. LTD.
4374/4B, Ansari Road, Daryaganj
New Delhi - 110 002
Ph.: 23261597, 23278000
Visit us at: www.anmolpublications.com

Encyclopaedic Dictionary of Chemistry

First Published, 1990

Reprint 1991, 1992, 1993, 1994, 1995, 1996, 1997, 1998, 1999

Reprint, 2004

ISBN 81-7041-274-9 (Set)

PRINTED IN INDIA

Published by J.L. Kumar for Anmol Publications Pvt. Ltd., New Delhi - 110 002 and Printed at Mehra Offset Press, Delhi.

Preface

This Dictionary, the first of its kind in India, is an effort to keep pace with continuing rapid developments and expanding vocabularies in Physical Chemistry. Throughout the process of compilation and editing this book our sincere effort was to make the dictionary serve as a ready reference for undergraduate and postgraduate chemistry students and for those appearing for competitive examinations, research scholars, teachers, and authors.

Every effort has been carefully made to write the entries in a clear and lucid style to provide both straight forward definitions and invaluable background information. At certain appropriate places, the line diagrams have been included whenever the meaning of a word can be best understood by means of a diagram.

The presentation throughout has been aimed at sustaining the interest of readers while enriching their vocabulary and comprehension of technical terms and expressions.

The dictionary will be of value to students of chemistry and to scientists and engineers working on science projects.

All comments from users on omissions of shortcomings will be most welcome.

Editors

Preface

A

Ab-Initio Calculations. Quantum mechanical calculations using the Schrodinger wave equation, which do not rely upon any experimental data for solution.

A. Abbreviation for absolute temperature.

Absolute. 1. Not dependent on or relative to anything else, *e.g.*, absolute zero. 2. Denoting a temperature measured on an *absolute scale*, a scale of temperature based on absolute zero. The usual absolute scale now is that of thermodynamic temperature ; its unit, the kelvin, was formerly called the degree absolute (°A) and is the same size as the degree Celsius. In British engineering practice an absolute scale with Fahrenheit-size degrees has been used : this is the Rankine scale.

Absolute Temperature. Symbol : *T*. A temperature which is defined by the relationship :

$$T=\theta+273.15$$

where θ refers to the Celsius temperature. The absolute scale of temperature was a fundamental scale based on Charles' law applied to an ideal gas :

$$V=V_0(1+\alpha\theta)$$

where V refers to the volume at temperature θ, V_0 the volume at 0, and α the thermal expansivity of the gas. At low pressures (when real gases show ideal behaviour) α has the value 1/273.15. Therefore, at $0=-273.15$ the volume of the gas theoretically becomes zero. In practice, of course, substances become solids at these temperatures. However, the extrapolation can be used for a scale of temperature on which—273.15°C corresponds to 0° (absolute zero). The scale is also called the *ideal-gas scale* ; on it temperature intervals were called *degrees absolute* (°A) or *degrees Kelvin* (°K), and were equal to the Celsius degree. It can be shown that the absolute temperature scale has been identical to the thermodynamic temperature scale (on which the unit is the kelvin).

Absolute Zero. The zero value of thermodynamic temperature ; 0 kelvin or −273.15°C.

Absorbent. (1) Any substance exhibiting the property of a absorption, *e g.*, absorbent cotton, so made by removal of the natural waxes present. See absorption (1). (2) A material that does not transmit certain wavelengths of incident radiation. See absorption (2).

Absorptiometer. An instrument which is used to measure the absorption of light by a liquid. Also the name of the apparatus which is used to determine the solubility of a gas in a liquid.

Absorption Bands. See absorption of light.

Absorption. Refers to a process in which a gas is taken up by a liquid or solid, or in which a liquid is taken up by a solid. In absorption, the substance absorbed goes into the bulk of the material. Solids that absorb gases or liquids often have a porous structure. The absorption of gases in solids is sometimes called *sorption.*

Absorption Bands. See absorption of light.

Absorption Coefficient of a Gas. The volume of gas measured at 0°C and 760 mm pressure which will dissolve in 1 ml of a liquid. The absorption coefficients in water at 0°C for several common gases are as follows : N_2, 0.024 ; O_2, 0.049 ; C_2H_4, 0.25 ; CO_2, 1.71 ; H_2S, 4.68 ; SO_2, 79.8 ; HCl, 506 ; NH_3, 1300.

Absorption Coefficient of Light. See Lambert's law and Beer's law.

Absorption of Light. Wnen light is incident on the surface of a transparent substanee part of the light gets reflected, the remainder is transmitted unchanged. If, however, the light is incident upon the surface of a black substance, *e.g.* lampblack, it is neither reflected nor transmitted ; all wavelengths get absorbed in a general absorption process. Many substances appear coloured because they have absorbed selectively all the wavelengths of white light except those corresponding to the particular colour which they appear. If the spectrum of the light transmitted by a coloured substance has been axamined it is found that certain wavelengths, called *absorption bands*, have been missing. For gases in the atomic state the spectrum of the transmitted light shows dark lines rather than bands. These *absorption lines* correspond to the wavelengths of light absorbed by the atoms.

Relationships between the intensity of incident light, sample thickness, concentration and intensity of transmitted light get embodied in Beer's law and Lambert's law.

Absorption Spectroscopy. Refers to any spectroscopic technique, qualitative or quantitative, which is depending upon the measurement of an absorption spectrum.

Absorption Spectrum. See spectrum.

Abundance. 1. The relative amount of a given element amongst others ; for example, the abundance of oxygen in the Earth's crust is approximately 50% by mass.

2. The amount of a nuclide (stable or radioactive) relative to other nuclides of the same element in a given sample. The *natural abundance* is the abundance of a nuclide as it occurs naturally. For instance, chlorine has two stable isotopes of masses 35 and 37. The abundance of ^{35}Cl is 75.5% and that of ^{37}Cl is 24.5%. For some elements the abundance of a particular nuclide depends on the source.

Accelerator. A catalyst which is added to increase the rate of a reaction.

Acid. An acid on the aqueous system may be defined as a substance which is capable of forming hydrogen ions when dissolved in water. Most inorganic acids may be considered as a compound of an acidic oxide and water ; where the oxide concerned is that of a metal, that oxide may exhibit amphoteric character, that is act sometimes as an acid and sometimes as a base. Aqueous solutions of acids are having a sharp taste, turn litmus red, liberate carbon dioxide from a metallic carbonate and give reactions characteristic of the anion present.

As free protons cannot exist, acidic properties can only be shown when the solvent can act as a proton acceptor, *i.e.*, as a base. Thus aqueous solutions of acids have the hydroxonium ion, H_3O^+. Acids can also exist in non-aqueous solvents. Since ammonia can also solvate a proton to give the ammonium ion, NH_4^+, substances which dissolve in ammonia to yield the ammonium ion, *e.g.*, NH_4Cl, are acids in that system.

Liquid water is ionized

$$2H_2O \rightleftharpoons H_3O^+ + OH^-$$

this ionization being the reverse of the neutralization reaction in water, substances giving hydroxyl ions are bases in water. Liquid ammonia is ionized and amides are bases on this system.

$$2NH_3 \rightleftharpoons NH_4^+ + NH_2^-$$

The concept of acids and bases has been extended to solvents which are ionized and yet do not have hydrogen : a substance giving the appropriate positive ion is an acid on that system. Thus bromine trifluoride ionizes

$$2BrF_3 \rightleftharpoons BrF_2^+ + BrF_4^-$$

and a substance giving the BrF_2^+ ion in solution, *e.g.*, $BrF_3.SbF_5$, is an acid in the system.

Typical organic acids have the $-C(O)OH$ group, but many other acid groupings, *e.g.*, the sulphonic $-S(O)_2OH$ give acidic properties to organic compounds. Phenols are having acidic properties and have been classified with enols as pseudo-acids.

The term acid was extended by Lewis to include substances which are electron acceptors. Thus $AlCl_3$ can accept electrons from a chloride ion forming the $[AlCl_4]^-$ ion and is a Lewis acid.

The 'strength' of an acid may be measured by the value of its dissociation constant, 'strong' acids, *e.g.*, HCl, HNO_3, being substantially fully ionized in solution and 'weak' acids predominately unionized.

Acidic. Having a tendency to release a proton or to accept an electron pair from a donor. In aqueous solutions the pH is a measure of the acidity, *i.e.*, an acidic solution is one in which the concentration of H_3O^+ exceeds that in pure water at the same temperature ; *i.e* , the pH is lower than 7.

Accumulator (**Secondary Cell** ; **Storage Battery**). A type of voltaic cell or battery that can be recharged by passing a current through it from an external d.c. supply. The charging current, which is passed in the opposite direction to that in which the cell supplies current, reverses the chemical reactions in the cell. The common types are the lead—acid accumulator and the nickel-iron accumulator.

Actinometer. Refers to a device which determines the intensity of a beam of radiation.

Actinon. An obsolete name for radon.

Activated Adsorption. There have been two types of adsorption ; in one the association is physical and in the other chemical. Physical adsorption has been the more common and involves van der Waals forces of attraction between the two phases. Chemical adsorption (chemisorption) involves the formation of chemical bonds between the two phases. Since chemisorption often has an activation energy associated with it, it is sometimes termed as activated adsorption.

Activated Complex. Refers to the partially bonded system of atoms in the transition state of a chemical reaction.

Activated Molecule. Molecules will only react on collision when they are having more than a certain minimum amount of energy. A molecule which has acquired more energy than the average amount possessed by other molecules, and is therefore in a more reactive condition, has been said to be activated. Molecules are activated when they absorb a quantum of light, or after coming in contact with a hot surface.

Activation Energy. Symbol E. The minimum energy a particle, molecule, etc., must acquire before it can react ; *i.e.*, the energy required to *initiate* a reaction regardless of whether the reaction has been exothermic or endothermic. Activation energy is often represented as an energy barrier that must be overcome if a reaction is to take place.

The activation energy (E_a) may be derived from the temperature dependence of the reaction rate using the Arrhenius equation.

Active Carbon. Carbon (charcoal) treated at high temperature with steam, air or CO_2. the process removing hydrogen from the surface and also opening up capillaries. It has been an excellent adsorbent for removal of small traces of impurities from a gas or liquid and is very specific the adsorption also taking place at low relative pressures. It is used in water and waste water treatment, air pollution control, as a catalyst, sugar refining, purification of chemicals and gases (gas masks), dry cleaning, rubber reclamation, cigarette filters.

Active Centres. In heterogeneous catalysis it is generally agreed that the chemisorption of at least one of the reactants has been a necessary prerequisite to catalysis. In 1925 H.S. Taylor proposed that the amount of the catalyst surface which was active for catalysis depended upon the reaction which was being catalysed, and that the adsorption occurred at active centres on the surface of the catalyst. Evidence for the existence of active centres has been obtained from the observation that the adsorption of small amounts of foreign substances, much less than that required for mono-layer coverage, can often completely poison the catalyst. The concept of active centres has been extended to enzymes and bacterial action.

Active Mass. An old term which was used to express the influence of concentration of a substance on the position of equilibrium in a reversible chemical reaction (see mass action, Law of). The active mass may be represented exactly by the thermodynamic activity.

Activity. Refers to a thermodynamic quantity which measures the effective concentration or intensity of a particular substance in a given chemical system. The *absolute activity*, a°, of a substance is given by $\mu = RT \ln a^\circ$, where μ is the chemical potential of the substance, R is the gas constant, T the absolute temperature. The *relative activity*, a, has been given by $\mu = \mu^\circ + RT \ln a$; μ° being the chemical potential of the substance in its standard state. For dilute, ideal solutions the activity has been directly proportional to the concentration ; for ideal gases, activity is proportional to the partial pressure of the gas.

Activity Coefficient (f or γ). Refers to a dimensionless factor by which the concentration (c) of a substance must be multiplied to give an exact measure of its thermodynamic activity (a) in a chemical system, *i.e.*, $a = fc$. It may be a measure of the deviation from ideal behaviour of the solution ; f is unity for an ideal mixture and greater or less than unity for a non-ideal system. Activity coefficients of ***electrolytes*** are regarded to be the geometrical mean of the single ion activities, the latter being hypothetical quantities which cannot be determined separately.

Activity Series. The elements arranged in order of their electrode potentials.

Additive Volumes, Law of. The vólume occupied by a mixture of gases is equal to the sum of the volumes which would by occupied by the constituents under the same conditions of temperature and pressure.

Adiabatic Change. Refers to a change during which no energy enters or leaves the system.

In an adiabatic expansion of a gas, mechanical work is done by the gas as its volume increases and the gas temperature falls. For an ideal gas undergoing a reversible adiabatic change it can be proved that

$$pV^{\gamma} = K_1$$

$$T^{\gamma} p^{1-\gamma} = K_2$$

and $$TV^{\gamma-1} = K_3$$

where K_1, K_2, and K_3 are constants and γ is the ratio of the principal specific heat capacities.

Adsorbate. Refers to a substance that is adsorbed on a surface.

Adsorption. Refers to a process in which a layer of atoms or molecules of one substance forms on the surface of a solid or liquid. All solid surfaces take up layers of gas from the surrounding atmosphere. The adsorbed layer may be held by chemical bonds (***chemisorption***) or by weaker van der Waals forces (***physisorption***).

Adsorption may in principle occur at all surfaces ; its magnitude has been particularly noticeable when porous solids, which have a high surface area, such as silica gel or charcoal are contacted with gases or liquids. Adsorption processes may be involving either simple unimolecular adsorbate layers or multilayers ; the forces which bind the adsorbate to the surface may be physical or chemical in nature.

Adsorption has been of technical importance in processes such as the purification of materials, drying of gases, control of factory effluents, production of high vacua, etc. Adsorption phenomena form the basis of heterogeneous catalysis and colloidal and emulsification behaviour.

Adsorption, Physical. Many adsorption processes have been not chemically specific and are readily reversible. The forces of attraction between the adsorbate and adsorbent have been weak and similar in nature to those responsible for the cohesion of molecules in the liquid state, namely, van der Waal's forces. For this reason physical adsorption often involves the formation of multiple layers at the adsorbent surface.

Aerosol, Aerogel. Refers to a dispersion in which a finely divided solid gets suspended in a gas is common, and is exemplified by smoke. When the dispersion medium has been air and the particles have been of colloidal dimensions, the system is termed an aerosol. Aerosols can often be formed by the rapid condensation of a vapour such as that of a metal oxide.

Affinity, Chemical. The chemical affinity of one substance for another is a measure of the tendency of the two substances to enter into chemical combination. The concept of affinity dates from the early days of chemistry. The term 'affinity of a chemical reaction has been used to denote the change of free energy in the system when the reaction takes place ; this is a true measure of the tendency to react.

Allotrope. See allotropy.

Allotropy. Refers to the ability of certain elements to exist in more than one physical form. The various forms are termed as allotropes. Sulphur and phosphorus have been the most common examples. Allotropy has been more common in groups IV, V, and VI of the periodic table than in other groups.

The difference in allotropes may be in bonding, *e.g.*, graphite and diamond, ozone and dioxygen, or in crystal form, *e.g.*, most metallic elements. The various forms are termed allotropes.

Alpha Particle, α-Ray. Refers to a He^{2+} ion, *i.e.*, a helium nucleus having no outer electrons. α-Particles are emitted with high velocity (*c.* 2×10^9 cm sec^{-1}) by Ra and other radioactive substances; they can travel for distances of several cm in air and other gases at normal pressures, causing these gases temporarily to become feeble conductors of electricity ('ionization'). Collision of α-particles with zinc sulphide or other substances brings about flashes of light, which enable individual particles to be counted (Crookes's 'spinthariscope').

α-Particles may be used, either with or without acceleration, as bombarding agents in nuclear disintegration reactions.

The energy, and thus the range, of α-particles, has been characteristic of the source of emission.

Ampere Symbol : A. The SI base unit of electric current which may be defined as the constant current that, maintained in two straight parallel infinite conductors of negligible circular cross section kept one metre apart in vacuum, would produce a force between the conductors of 2×10^{-7} newton per metre.

Ampholyte Ion. See zwitterion.

Amphoteric. A material that is able to display both acidic and basic properties. The term has been most commonly applied to the oxides and hydroxides of metals that can form both cations and complex anions. For example, zinc oxide dissolves in acids to yield zinc salts and also dissolves in alkalis to yield zincates, $[Zn(OH)_4]^{2-}$.

Andrews' Experiment. The experiment used for an investigation (1861) into the relationship between pressure and volume for a mass of carbon dioxide at constant temperature. The resulting isothermals showed clearly the existence of a critical point and led to greater understanding of the liquefaction of gases.

Angle of Contact. A liquid will either rise in a glass capillary tube (*e g.*, water) or be depressed (*e.g.*, mercury) depending on the relative strengths of the intermolecular forces in the different phases—in particular on the cohesive forces within the liquid and the adhesive forces between the solid and the liquid. If water rises in a capillary tube the angle which the liquid surface makes with the tube walls is less than 90° and a concave meniscus results. Water is then said to wet the glass surface. In the figure the angle α between the liquid surface and the solid has been the angle of contact. If α has been less than 90° a concave meniscus results and if α is greater than 90° the meniscus is convex. When $\alpha = 90°$ a plane meniscus is given. The angle of contact of water on pure paraffin wax has been about 110°.

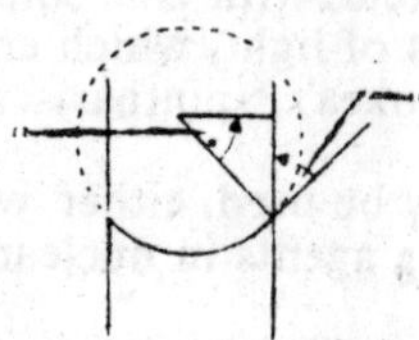

Used as a measure of the wetting characteristics of a liquid/solid system. See surface tension wetting agents. The presence of surfactants, *e.g.*, detergents, lowers the surface tension of the liquid and, in consequence, lowers the contact angle between that liquid and a solid surface.

Amu, Atomic Mass Units. Units of atomic weight in terms of an arbitrary standard (^{12}C).

Angstrom Unit, A. A unit of length which is equivalent to 10^{-8} cm, (10^{-10} m). It is used to describe molecular and lattice dimensions and wavelengths of visible light.

Anharmonicity. Harmonic vibrations have been characterized by a restoring force which is proportional to the displacement. While at very small amplitudes the vibrations of the atomic nuclei in molecules have been harmonic, band spectra reveal that at larger amplitudes, especially those approaching the dissociation point of the bond, the vibrations are markedly anharmonic. The deviations from harmonic behaviour have been expressed by the anharmonicity constant.

Anion. A negatively charged ion which is formed by addition of electrons to atoms or molecules. In electrolysis anions get attracted to the positive electrode (the anode). *Compare* cation.

Anisotropic. A substance is said to be anisotropic when any of its physical properties (*e.g.*, thermal or electrical conductivity, refractive index) have been different in different principal directions. Crystals, other than those belonging to the cubic system, are anisotropic. Certain substances (*e.g*, *p*-azoxyanisole) melt to form a cloudy anisotropic liquid ('liquid crystals'), which again 'melts' at a higher temperature to a normal (isotropic) liquid.

Anode. Refers to the electrode which carries the positive charge in an electrochemical cell. In electrolysis the anode has been the electrode at which the negative ions get discharged.

Anodic Oxidation. Refers to the process of oxidation in solution by means of an electric current.

Anodizing. Process of anodic oxidation of metals with the object of giving protection against corrosion. Commonly used for aluminium sheet which is made the anode of a sulphuric-chromate (VI) bath, and becomes coated with a layer of porous oxide. The pores are sealed by dipping in hot water ; the layer may be coloured by adsorbed dyes.

Antiferromagnetism. Refers to the Magnetic behaviour in which the magnetic susceptibility drops with decreasing temperature. There has been generally a characteristic temperature, the Neel point, above which magnetic behaviour is normal. The behaviour is discussed in terms of two interpenetrating lattices within the crystal with opposed magnetic moments. The two lattices can, in some cases, completely cancel out the magnetism.

Anti-Fluorite Structure. Adopted by compounds M_2X with the cations occupying the F^- positions of the fluorite structure and the anions occupying the Ca^{2+} positions. Adopted by *e.g.*, K_2S and K_2O.

Anti-Cathode. See X-ray tube.

Anti-Foaming Agents. Refer to substances which prevent the formation of foams. They get strongly adsorbed by the liquid medium but they do not have the electrical or mechanical properties required to form a foam. Examples of anti-foaming agents have been polyamides which have been used in boiler-feed water and octanol used in electroplating baths and in paper making. Low concentrations of silicones also find quite general use.

Anti-Foggants. Materials which are added to emulsions (5-nitrobenzimidazole, KBr) or developers to prevent fogging of photographic materials, during storage or development.

Arc Spectrum. The emission spectrum obtained when a substance gets excited by application of an electric are rather than an electric spark which gives a *spark spectrum*. As the spark brings about a greater degree of excitation of the atoms than does the arc, the spark spectrum contains more lines and lines which are enhanced in intensity compared with the arc spectrum.

Arrhenius Equation. An equation which is relating the rate constant of a chemical reaction and the temperature at which the reaction is occurring :

$$k = A\exp(-E/RT)$$

where A is a constant, k the rate constant, T the thermodynamic temperature in kelvins, R the gas constant, and E the activation energy of the reaction.

Reactions proceed at different rates at different temperatures, *i.e.*, the magnitude of the rate constant is temperature dependent. The arrhenius equation is often written in a logarithmic form, *i.e.*,

$$\log_e k = \log_e A - E/2.3RT$$

This equation enables the activation energy for a reaction to be determined.

The pre-exponential factor A is a collision frequency factor, which is comparatively insensitive to temperature. $A = pZ$, where Z is the collision frequency and p is a 'steric factor', which is a measure of the efficiency of the collision for effective reaction. The Arrhenius equation applies to processes such as diffusion, electrolytic conduction, viscous flow, etc., as well as chemical reactions.

Arrbenius Theory. See ionisation theory.

Atmolysis. Refers to the separation of a mixture of gases by means of their different rates of diffusion. Usually, separation is achieved by allowing the gases to diffuse through the walls of a porous partition or membrane.

Atmosphere 1. **(Atm.)**. A unit of pressure equal to 101,325 pascals. This is equal to 760.0 mmHg. The actual atmospheric pressure fluctuates around this value. The unit is usually used for expressing pressures well in excess of standard atmospheric pressure, *e.g.*, in high-pressure chemical processes. 2. See earth's atmosphere.

Atmospheric Pressure. The pressure exerted by the weight of the air above it at any point on the earth's surface. At sea level the atmosphere will support a column of mercury about 760 mm high. This decreases with increasing altitude. The standard value for the atmospheric pressure at sea level in SI units is 101,325 pascals.

Atomic Energy. Mass and energy have been equivalent, $E=mc^2$ (E, energy in ergs : *m*, mass in grams ; *c*, velocity of light= 3×10^{10} cm/s^{-1}). In the building up of nuclei, part of the mass of the constituent particles appears as binding energy ; the ratio (mass defect) : (atomic number) is termed as the packing fraction. The elements of low atomic numbers and those of very high atomic number are having a negative packing fraction ; those of intermediate atomic number have a positive packing fraction and are more stable. If heavy nuclei are split to form nuclei of the more stable elements there is a loss of mass ; there is a similar loss of mass if light nuclei are fused to give intermediate elements. This loss in mass is evolved as energy.

Atomic Heat. The atomic heat of an element may be defined as the product of its atomic weight and specific heat. It is the heat capacity, water being taken as unity, of one gram-atom of the element. See Dulong and Petit's Law.

Atomic Number. Refers to the number of unit positive charges (protons) contained in the nucleus of an atom of the element. It is the property of an element which determines the position of the element in the periodic table. The atomic number can be determined directly from the X-ray spectrum of the element.

Atomic Spectrum. See line spectrum.

Atomic Weights, At. Wt. Defined relative to ^{12}C as 12.0000, atomic weights are ratios of masses and are not to be confused with the actual mass of an atom. Also termed relative atomic mass.

Atto- Symbol : A. A prefix denoting 10^{-18}. For example, 1 attometre (am) $= 10^{-18}$ metre (m).

Auger Spectroscopy. The photoinduced ejection of an electron from an atom causes the creation of an electron hole or vacancy. This process is followed by an internal electronic reorganization in which an electron from a higher energy level falls into the vacancy. The excess energy associated with this process is released either as X-ray fluorescence, or by the ejection of a second electron (the auger electron). The kinetic energy of the auger electron (E_A) emitted from the L shell is given by

$$E_E = (E_K - E_L) - E_L$$

where E_K and E_L are the binding energies of electrons in the K and L shells. Auger spectroscopy is particularly useful with the lighter elements and may be used to identify elements in the surface layers of solid samples.

Autocatalyst. See catalyst.

Auto-Catalysis. The process whereby the reaction products catalyse the further reaction of the reactants.

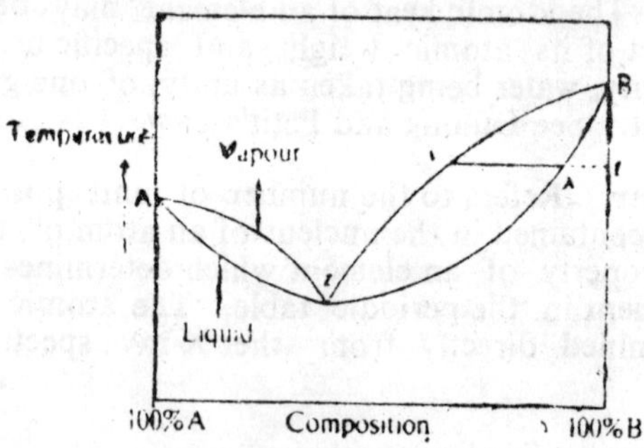

Avogadro Constant. Symbol : N_A. The number of particles in one mole of a substance. Its value is 6.02252×10^{23} mol^{-1}.

Avogadro's Law. The principle that equal volumes of all gases at the same temperature and pressure contain equal numbers of molecules. It is often called *Avogadro's hypothesis*. It is strictly true only for ideal gases.

Avogadro's Number. L The number of particles (atoms or molecules) in one mole of any pure substance. $L=6.023\times10^{23}$. It has been determined by many methods including measurements of Brownian movement, electronic charge and the counting of α particles.

Axes of Symmetry. See symmetry elements.

Axial Ratios. Refers to the ratio of the cell dimensions (a, b and c) in a crystal ; b being taken as unity. In triclinic, monoclinic and orthorhombic systems

$$a\neq b\neq c$$

the axial ratios have been, therefore, quoted as

$$a/b : 1 : c/b$$

In hexagonal and tetragonal systems

$$a=b\neq c$$

only $c : a$ is required. In rhombohedral and cubic systems $a=b=c$. See crystal structure and crystal systems.

Axis of Symmetry. See crystal symmetry and symmetry elements.

Azeotrope (Azeotropic Mixture). Refers to a mixture of liquids for which the vapour phase has the same composition as the liquid phase. It therefore boils without change in composition and consequently without progressive change in boiling point.

The composition and boiling points of azeotropes have been found to vary with pressure, indicating that they are not chemical compounds. Azeotropes may be broken by distillation in the presence of a third liquid, by chemical reactions, adsorption, or fractional crystallization.

Azeotropic Distillation. A method used to separate mixtures of liquids that cannot be separate by simple distillation. Such a mixture is called an azeotrope. A solvent is added to form a new azeotrope with one of the components, and this is then removed and subsequently separated in a second column. An example of the use of azeotropic distillation is the dehydration of 96% ethanol to absolute ethanol using benzene. Azeotropic distillation is not widely used because of the difficulty of finding inexpensive non-toxic non-corrosive solvents that can be easily removed from the new azeotrope.

Azeotropic Mixtures. Mixtures of liquids when distilled may reach stage at which the composition of the liquid is the same as that of the vapour.

Many liquid mixtures exhibit a minimum boiling point (*e.g.*, methanol and chloroform ; *n*-propanol and water) whilst others show a maximum b.p. (*e.g.* hydrochloric acid and water) when vapour pressures are plotted as a function of change in composition of the mixture. An example of a system (A+B) where a minimum is seen is shown in the figure. The upper curve represents the composition of the vapour phase whilst the lower that of the liquid. When the liquid of composition *x* boils at temperature *t* it is seen that the composition of the vapour (*v*) differs in that it is richer in constituent A than the liquid, and an increase in the b.p. would eventually give pure B.

At *z* in the curve, however (the minimum of vapour pressure), the solution and vapour are in equilibrium and the liquid at this point will distil without any change in composition. The mixture at *z* is thus said to be azeotropic or a constant boiling mixture. It should be noted, however, that the composition of the azeotropic mixture does vary with pressure.

(DPC—1)

B

Babo's Law. The vapour pressure of a liquid is decreased when a solute is added, the amount of the decrease being proportional to the amount of solute dissolved. The law was discovered in 1847 by the German Chemist Lambert Babo (1818-99).

Back e.m.f. Refers to an e.m.f. that opposes the normal flow of electric charge in a circuit or circuit element. In some electrolytic cells a back e.m.f. is brought about by the layer of hydrozen bubbles that builds up on the cathode as hydrogen ions pick up electrons and form gas molecules (*i.e.* as a result of polarization of the electrode).

Background Radiation. Low intensity ionizing radiation present on the surface of the earth and in the atmosphere as a result of cosmic radiation and the presence of radio-isotopes in the earth's rocks, soil, and atmosphere. The radioisotopes are either natural or the result of nuclear fallout of waste gas from power stations Background counts must be taken into account when measuring the radiation produced by a specified source.

Band Spectrum. Refers to a spectrum that appears as a number of bands of emitted or absorbed radiation. Band spectra have been characteristic of molecules. Often each band can be resolved into a number of closely spaced lines. The bands correspond to changes of electron orbit in the molecules. The close lines seen under higher resolution are the result of different vibrational states of the molecule.

Bar. A unit of pressure which may be defined as 10^5 pascals. The millibar (mb) is more common ; it is used for measuring atmospheric pressure in meterology.

Barn. In a nuclear reaction, the probability of reaction may be expressed in terms of the nuclear cross section, σ. If N target atoms per cm^3 are bombarded with I_o particles, the number of particles transmitted is given by the equation

$$I=I_o e^{-N\sigma x}$$

where x is the thickness of the target sample. The cross section of the target atoms σ is expressed in barn units ; 1 barn=10^{-24} cm^2.

Barometric Condenser. A plant for condensing steam at pressures below atmospheric. It consists of a vassel situated at the top of a vertical tube called a *barometric leg*. Steam passing into the vessel is condensed by direct contact with a spray of cold water and discharges by gravity through the leg. The water in the leg acts as a seal which enables the vacuum to be maintained.

Barometric condensers are cheap and economical to operate. They are commonly used in conjunction with evaporators.

Base. On the simplest view a base may be defined as a substance which in aqueous solution reacts with an act to form a salt and water only, *i.e.* is a substance which furnishes hydroxyl ions. A more general definition (Lowry, Bronsted), which also applies to non-aqueous solutions, states that a base is a substance with a tendency to gain protons. By this definition, OH^- and the anions of weak acids, *e.g.* CH_3COO^-, are bases in aqueous solution.

For ionized solvents which do not contain protons a base is a substance which reacts with one acid of that system to give a salt and the solvent. Thus the base $KBrF_4$ reacts with the acid Br_2SbF_6 to give the salt $KSbF_6$ and BrF_3 in bromine trifluoride.

A Lewis base may any molecule with 'excess' electrons. Thus ammonia, with a lone pair of electrons, is a Lewis base.

As an adjective applied to metals base represents the opposite of noble, *i.e.* a base metal would be attacked by mineral acids.

Base-catalysed Reaction. A reaction which is catalysed by bases. Typical base catalysed reactions are the Claisen condensation and the aldol reaction, in which the first step is abstraction of a proton to give a carbanion.

Base Dissociation Constant. See dissociation.

Base Unit. A unit that is defined in terms of reproducible physical phenomena or prototypes, rather than of other units. The metre, for example, is a base unit in the SI, being defined in terms of the wavelength of a particular light emission.

Basic. Having a tendency to release OH^- ions. Thus any solution in which the concentration of OH^- ions is greater than that in pure water at the same temperature is described as basic ; *i.e.* the pH is greater than 7.

Basicity Constant. See dissociation.

Basic Salts. Compounds intermediate between simple salts and oxides or hydroxides. Formed by hydrolysis of metal salts and often precipitated by sodium carbonate solution.

The term is often restricted to hydroxyhalides (such as $Pb(OH)Cl$, $Mg_2(OH)_3Cl_34H_2O$, and $Zn(OH))F$ and hydroxy-oxy salts (for example, $2PbCO_3.Pb(OH)_2$, $Cu(OH)_2.CuCl_2$,, $Cu(OH)_2.CuCO_3$). The hydroxyhalides are essentially closely packed assemblies of OH^- with metal or halide ions in the octahedral holes. The hydroxy-oxy salts have more complex structures than those implied by the formulae.

Battery. A number of electric cells joined together. The common car battery, or accumulator, usually consists of six secondary cells connected in series to give a total e.m.f. of 12 volts. A torch battery is usually a dry version of the Leclanche primary cell, two of which are often connected in series. Batteries may also have cells connected in parallel, in which case they have the same e.m.f. as a single cell, but their capacity is increased, *i.e.* they will provide more total charge. The capacity of a battery is usually specified in ampere-hours, the ability to supply 1 A for 1 hr, or the equivalent.

B.C.C. Body-centred cubic crystal.

Beckmann Thermometer. A thermometer for measuring small changes of temperature. It consists of a mercury-in-glass thermometer with a scale covering only 5 or 6°C calibrated in hundredths of a degree. It has two mercury bulbs, the range of temperature to be measured is varied by running mercury from the upper bulb into the large lower bulb. It is used particularly for measuring depression of freezing point or elevation of boiling point of liquids when solute is added, in

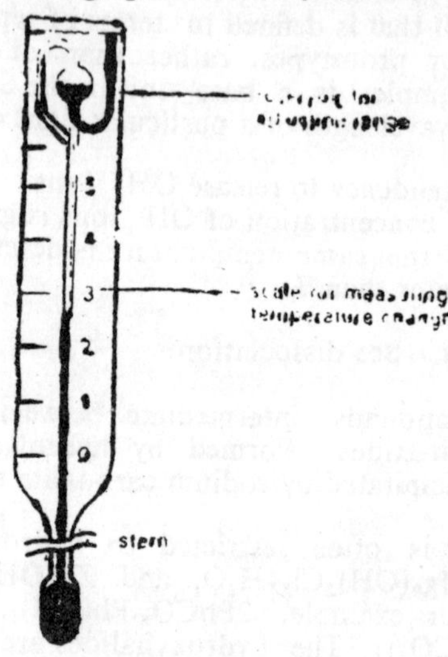

Beckmann thermometer

order to find relative molecular masses. The instrument was invented by the German chemist E. O. Backmann (1853-1923).

Beer's Law. This states that the proportion of light absorbed depends on the Thickness (d) of the absorbing layer, and on the *molecular concentration* (c) of the absorbing substance in in the layer. It is an extension of Lambert's law, and may be written in the form

$$I = I_o e^{-Kcd},$$

where I_o is the intensity of the incident radiation, and I that of the transmitted radiation; and K is a constant, the *molecular absorption coefficient*, which is characteristic of the absorbing substance for light of a given frequency. The *molecular extinction coefficient*, a_m, is the thickness is cm of a layer of a molar solution which reduces the intensity of light passing through it to one-tenth of its original value, a_m,$=0.4343$ K. The law does not hold for substances whose absorption spectra and the species present are affected by concentration.

Bentonite. A clay-like mineral eonsisting largely of montomorillonite. Forms gelatinous suspensions at low concentrations and used as a binder for foundry sand and in oil-drilling needs. One form has been used for its absorptive properties.

Beta Particle, Beta Ray. Refers to an electron which gets emitted during certain radioactive disintegrations. β-Particles, as normally emitted, possess high velocity, sometimes comparable with that of light, and a long path length in air at atmospheric pressure. They may be deflected by both electrostatic and magnetic fields, detected in radioactive counters.

Betatron. An appartus which is used for accelerating electrons. Operates by producing an electric force around the axis of a fluctuating magnetic field to produce energies of the order of up to 2×10^9 electron volts, although energy losses by radiation are high.

Bimolecular. Describing a reation or step that involves two molecules. For example, the decomposition of hydrogen iodide,

$$2HI \rightarrow H_2 + I_2$$

takes place between two molecules and is therefore a bimolecular reaction. Other common examples are :

$$H_2O_2 + I_2 \rightarrow OH^- + HIO$$

$$OH^- + H^+ \rightarrow H_2O$$

All bimolecular reactions are second order, but some second-order reactions are not bimolecular.

Bimolecular Reaction. A step in a chemical reaction that involves two molecules. See molecularity.

Bioluminescence. The phenomenon of light emission by plant or animal organisms, often due to the reaction of oxygen with an oxidizable substrate (luciferin) catalysed by an enzyme (a luciferinase).

Birefringence. See double refraction.

Body-Centred Cubic Crystal (b.c c.). A crystal structure in which the unit cell has an atom, ion, or molecule at each corner of a cube and at the centre of the cube. In this type of structure the coordination number is 8. It is less close-packed than the face-centred cubic structure. The alkali metals form crystals with body-centred cubic structures. See also cubic crystal.

Body Centred Lattice. A lattice which is having a point (practically atom or molecule) at the corners of the cell and at the body centre. Body centred cubic lattices (bcc) are adopted by many metals.

Bohr Magneton. Refers to the unit of magnetic moment. The theoretical values of magnetic moments for unpaired electrons, taking spin only into account are : 1=1.73 B.M., 2=2.83, 3=3.87, 4=4.90, 5=5.92, 6=6.93, 7=7.94. The observed values may be altered by metal-metal bonding, ferromagnetism*, anti-ferro-magnetism*, spin-orbit coupling and other effects.

Bohr Theory. A theory which was introduced by Niels Bohr (1911) to explain the spectrum of atomic hydrogen. The model he used was that of a nucleus with charge $+e$, orbited by an electron with charge $-e$, moving in a circle of radius r. If v has been the velocity of the electron, the centripetal force, mv^2/r is equal to the force of electrostatic attraction, $e^2/4\pi\epsilon_0 r^2$. Using this, it can be proved that the total energy of the electron (kinetic and potential) is $-e^2/8\pi\epsilon_0 r$.

If the electron is considered to have wave properties, then there must be a whole number of wavelengths around the orbit, otherwise the wave would be a progressive wave. For this to occur

$$n\lambda=2\pi r$$

where n is an integer, 1, 2, 3, 4,...... The wavelength, λ, is h/mv, where h is the Planck constant and mv the momentum. Thus for a given orbit :

$$nh/2\pi=mvr$$

This implies that orbits are possible only when the angular momentum (mvr) is an integral number of units of $h/2\pi$. Angular momentum is thus quantized. In fact, Bohr in his theory did not use the wave behaviour of the electron to derive this relationship. He assumed from the beginning that angular momentum was quantized in this way. Using the above expressions it can be shown that the electron energy is given by

$$E = -me^4/8\epsilon_0^2n^2h^2$$

Different values of n (1, 2, 3, etc.) correspond to different orbits with different energies; n is the principal quantum number. In making a transition from an orbit n_1 to another orbit n_2 the energy difference $\triangle W$ is given by: $\triangle W = W_1 - W_2 =$

$$me^4(1/n_2^2 - 1/n_1^2)/8\epsilon_0^2h^2$$

This is equal to hv where v is the frequency of radiation emitted or absorbed. Since $v\lambda = c$, then

$$1/\lambda = me^4(1/n_1^2 - 1/n_2^2)/8\epsilon_0^2ch^3$$

The theory is in good agreement with experiment in predicting the wavelengths of lines in the hydrogen spectrum, although it is less successful for larger atoms. Different values of n_1 and n_2 correspond to different spectral series, with lines given by the expression :

$$1/\lambda = R(1/n_1^2 - 1/n_2^2)$$

R is the *Rydberg constant.* Its experimental value is 1,096 78 $\times 10^7$ m^{-1}. The value from Bohr theory ($me^4/8\pi\epsilon_1^2ch^2$ is 1.097 00 $\times 10^7$ m^{-1}.

Boiling Point (B.P.). Refer to the temperature at which the saturated vapour pressure of a liquid equals the external atmospheric pressure. As a consequence, bubbles form in the liquid and the temperature remains constant until all the liquid has evaporated. As the boiling point of a liquid depends on the external atmospheric pressure, boiling points are usually quoted for standard atmospheric pressure (760 mmHg= 101 325 Pa).

Boiling. Refers to the process by which a liquid gets converted into a gas or vapour by heating at its boiling point. At this temperature the vapour pressure of the liquid is equal to the external pressure, and bubbles of vapour can form within the liquid.

Boiling Point-Composition Diagram. A diagram for a two-component liquid system representing both the variation of the boiling point and the composition of the vapour phase as the liquid-phase composition gets varied. A mixture of A and B at composition L_1 would have a boiling point T_1 and a vapour

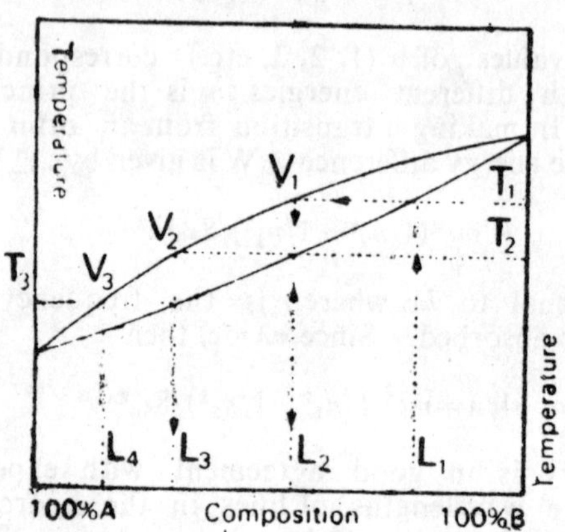

Boiling point—composition diagram

composition V_1, which when condensed would give a liquid at L_2. L_2 would boil at T_2 to give vapour V_2 equivalent to liquid of composition L_3, and so on. Thus the whole process of either distillation or fractionation of this system, will lead to progressive enrichment in component A.

For perfect solutions obeying Raoult's law the curves would coincide but in real cases there will be sufficient inter-molecular attraction to cause deviation from this. The separation of the curves as well as the difference in boiling points determines the performance of fractionation columns. See also constantboiling mixture.

Boiling-Point Diagram. A graph which shows equilibrium compositions of liquid and vapour plotted against temperature for a two-component system at any given pressure.

Boiling Point, Elevation of. Refers to the increase in boiling point of a solution, compared with that of the pure solvent, due to the dissolved substance. The elevation of boiling point is proportional to the weight of a particular solute. Molecular proportions of different solutes produce the same elevation. The following values represent the elevation of boiling point which is prdouced in different solvents by dissolving 1 mol of any solute in 100g of solvent. This is termed as the molecular elevation of boiling point.

Solvent	*Molecular elevation*
Water	5.2°
Chloroform	38.8°
Ether	21.1°
Acetone (propanone)	17.2°
Benzene	25.7°
Alcohol (ethanol)	11.5°

Boltzmann Constant. Symbol *k*. The ratio of the universal gas constant (R) to the Avogadro constant (N_A). It may be thought of therefore as the gas constant per molecule :

$$k = R/N_A = 1{,}380\ 622 \times 10^{-23}\ JK^{-1}$$

It is named after the Austrian physicist Ludwig Boltzmann (1844-1906).

Bomb Calorimeter. An apparatus used for measuring heats of combustion (*e.g.*, calorific values of fuels and foods). It consists of a strong container in which the sample is scaled with excess oxygen and ignited electrically. The heat of combustion at constant volume can be calculated from the resulting rise in temperature.

Bond Energy. The energy involved in forming a bond. For ammonia, for instance, the energy of the N-H bond is one-third of the energy involved for the process

$$NH_3 \rightarrow N + 3H$$

It is thus one third of the heat of atomization.

The *bond dissociation energy* is a different quantity to the bond energy. It is the energy required to break a particular bond in a compound, *e.g.*,

$$NH_3 \rightarrow NH_2 + H$$

Born-Haber Cycle. A thermodynamic cycle derived by application of Hess's law. Commonly used to calculate lattice energies of ionic solids and average bond energies of covalent compounds. *e.g.* NaCl :

$$\begin{array}{ccc} NaCl(s) & \xleftarrow{-U_0} & Na^+(g)+Cl^-(g) \\ \uparrow \Delta Hf^\circ & & \uparrow I \quad \uparrow -E \\ Na(s)+\frac{1}{2}Cl_2(g) & \xrightarrow[S+\frac{1}{2}D]{} & Na(g)+Cl(g) \end{array}$$

S = Heat of sublimation of sodium

D = Dissociation energy of chlorine

I = Ionization energy of sodium

E = Electron affinity of chlorine

U_0 = Lattice energy of sodium chloride

ΔHf° = Heat of formation of sodium chloride.

From the Born-Haber cycle it follows that

$$\Delta H_f^\circ = S+\tfrac{1}{2}D+I-E-U_0$$

All terms in the equation can be determined experimentally except U_0, which can thus be calculated.

Similar cycles may be drawn for covalent compounds. *e.g.* PCl_5 :

$$\begin{array}{ccc} P(s)+1/2Cl_2(g) & \xrightarrow{\Delta Hf^\circ} & PCl_5(g) \\ \downarrow S \quad \downarrow 1/2D & & \nearrow -5B \\ P(g)+5Cl(g) & & \end{array}$$

B = Average bond energy of P-Cl bond.

From the cycle it follows that :

$$\Delta Hf^\circ = S+\frac{5}{2}D-5B$$

Boundary Layer. If a fluid flows over a surface, due to the frictional resistance there is a layer of fluid in the immediate vicinity of the surface which has a velocity smaller than that in the bulk of the fluid. This is termed as the boundary layer. In the part of the layer nearest the surface flow is laminar, that is, the layers of fluid move parallel to each other and there is no mixing between them. Transfer of heat or material across the boundary layer can only take place by the relatively slow process of molecular diffusion and it is therefore of the greatest technical importance in that it determines the heat and mass transfer properties of the system.

Bowl Classifier. Refers to a device that separates solid particles in a mixture of solids and liquid into fractions according to particle size. Feed enters the centre of a shallow bowl, which contains revolving blades. The coarse solids collect on the bottom, fine solids at the periphery.

Boyle's Law. At a constant temperature, the pressure of a fixed mass of a gas is inversely proportional to its volume, *i.e* ,

$$pV=K$$

where K is a constant. The value of K depends on the temperature and on the nature of the gas. The law holds strictly only for ideal gases. Real gases follow Boyle's law at low pressures and high temperatures. See gas laws.

Although exact at low pressures, the law is not accurately obeyed at high pressures because of the finite size of molecules and the existence of intermolecular forces.

This law was discovered in 1602 by the Irish physicist Robert Boyle (1627-91). On the continent of Europe it is known as ***Mariotte's** law* after E. Mariotte (1620-84), who discovered it independently in 1676. See also gas laws.

Bragg Equation. When a beam of monochromatic X-rays of wavelength λ impinges on a crystal, strong scattering occurs in certain directions only this is the phenomenon of X-ray diffraction. For diffraction the path difference between waves scattered from successive planes of atoms in the crystal must equal an integral number of wavelengths, n (n=1,2,3,...). This condition is expressed by the equation

$$n\lambda=2d \sin \theta$$

where d is the distance separating successive planes in the crystal and θ is the angle which the incident beam of X-rays makes with the same planes.

Bragg Scattering. Coherent elastic scattering of monochromatic neutrons by a set of crystal planes.

Branched Chain. See chain.

Bremsstrahlung. See X-radiation.

British Thermal Unit (Btu). The Imperial unit of heat, being originally the heat required to raise the temperature of 1 lb of water by 1 °F. 1 Btu is now defined as 1055 06 joules.

Buffer Solutions. It is often desirable to prepare a solution of definite pH, made up in such a way that this pH alters only gradually with the addition of alkali or acid. Such a solution is called a buffer solution, and generally consists of a solution of a salt of a weak acid in the presence of the free acid itself, *e.g.* sodium ethanoate and ethanoic acid. The pH of the solution is determined by the dissociation equilibrium of the free acid :

$$\frac{[H^+][CH_3COO^-]}{[CH_3COOH]} = K$$

The sodium ethanoate which is largely dissociated, serves as a source of ethanoate ions, which combine with any hydrogen ions which may be added to the solution to yield more of the acid. The addition of hydrogen ions has therefore much less effect on such a solution than it would have on water. In a similar manner the solution of the salt of a strong acid and a weak base, in the presence of a weak base, has a pH that is insensitive to additions of alkali.

Bumping. Violent boiling of a liquid caused when bubbles form at pressure above atmospheric pressure.

Bunsen Cell. A primary cell consisting of a zinc cathode immersed in dilute sulphuric acid and a carbon anode immersed in concentrated nitric acid. The electrolytes are separated by a porous pot. The cell gives an e.m.f. of about 1.9 volts.

Burger's Vector. A measure of the crystal lattice displacement resulting from the passage of a dislocation.

C

Cadmium Cell. See Weston cell.

Caesium chloride Structure. Refers to a form of crystal structure that consists of alternate layers of caesium ions and chloride ions with the centre of the lattice occupied by a caesium ion in contact with eight chloride ions (*i.e.* four chloride ions in the plane above and four in the plane below).

Calcium Fluoride Structure. (Flourite structure). Refers to a form of crystal structure in which each calcium ion is surrounded by eight fluoride ions arranged at the corners of a cube and each fluoride ion is surrounded tetrahedrally by four calcium ions.

Calomel Half Cell (Calomel Electrode). A type of half cell in which the electrode is mercury coated with calomel (HgCl) and the electrolyte is a solution of potassium chloride and saturated calomel. The standard electrode potential is -0.2415 volt (25°C). In the calomel half cell the reactions are

$$HgCl(s) \rightleftharpoons Hg^+(aq) + Cl^-(aq)$$
$$Hg^+(aq) + e \rightleftharpoons Hg(s)$$

The overall reaction is

$$HgCl(s) + e \rightleftharpoons Hg(s) + Cl^-(aq)$$

This is equivalent to a $Cl_2(g) \mid Cl^-(aq)$ half cell in which the pressure is the dissociation pressure of HgCl.

The hydrogen electrode is not easy to use as a general standard for laboratory use, and it is often replaced by a calomel electrode which has a potential known in terms of that of the hydrogen electrode. Three types of calomel electrode have been used, *viz.* with 0.1m, 1.0m and staturated KCl. Their respective potentials in volts on the hydrogen scale at 298 K are : 0.3338, —0.2800, —0.2415.

For fuels it is quite usual to employ kilocalories per kilogram as a measure of calorific value while for foods the unit used is kilocalories per gram. This however, is often abbraviated to 'Calories', so that a value for carbohydrates of 4.1 'Calories' per gram is 4100 calories per gram.

Calorie (Kilogram Calorie ; Kilocalorie). 1000 calories. This unit is still in limited use in estimating the energy value of foods, but is obsolescent.

Calorific Value The heat per unit mass produced by complete combustion of a given substance. Calorific values are used to express the energy values of fuels ; usually these are expressed in megajoules per kilogram (MJ kg^{-1}). They are also used to measure the energy content of foodstuffs ; *i.e.* the energy produced when the food is oxidized in the body. The units here are kilojoules per gram (kJ g^{-1}), although Calories (kilocalories) are often still used in non-technical contexts. Calorific values are measured using a bomb calorimeter.

For hydrogen, and all hydrogen-containing fuels, there are two calorific values : *gross* or higher calorific values which assume that all the water formed is condensed and cooled to atmospheric temperature ; *net* or lower calorific values imply that none of the water is condensed.

Calorimeter. A device which is used for the measurement of calorific values of fuels and foods or for the measurement of heats of chemical reactions. For solid and liquid fuels the bomb calorimeter is used, while for gases a suitable gas calorimeter is used. Sophisticated micro-calorimeters are available for the accurate measurement of very small increases in temperature. These usually employ thermocouples or thermistors rather than thermometer.

Candela. Symbol : cd. The SI base unit of luminous intensity, defined as the intensity (in the perpendicular direction) of the black-body radiation from a surface of 1/600,000 square metre at the temperature of freezing plantinum and at a pressure of 101 325 pascals.

Capillary Condensation. If a substance wets the wall of a capillary tube in which it is contained, and possesses a concave meniscus, the vapour pressure above the meniscus is less than that in contact with a plane surface of the liquid at the same temperature. This is expressed by the equation

$$\log_\epsilon \frac{p}{p_s} = 2\frac{\sigma v}{r\mathrm{RT}}$$

(p is the pressure at the concave surface, p_s the normal vapour pressure at the same absolute temperature T, σ the surface tension, r the radius of the capillary tube, R the gas constant and v the mol. volume of the liquid). There is thus a considerable tendency for a vapour to condense in the fine capillaries of a porous material such as charcoal. Capillary condensation is certainly the mechanism of the sorption of water by silica gel, and probably also at high humidities by charcoal, but it is unsatisfactory as a general theory of physical adsorption.

Carbonation. The solution of carbon dioxide in a liquid under pressure, as in carbonated soft drinks.

Carbon Dating (Radiocarbon Dating). A method of estimating the ages of archaelogical specimens of biological origin. As a result of cosmic radiation a small number of atmospheric nitrogen nuclei are continuously being transformed by nutron bombardment into radioactive nuclei of carbon—14 :

$$^{14}_{7}\mathrm{N} + n \rightarrow {}^{14}_{6}\mathrm{C} + p$$

Some of these radiocarbon atoms find their way into living trees and other plants in the form of carbon dioxide, as a result of photosynthesis. When the tree is cut down photosynthesis stops and the ratio of radiocarbon atoms to stable carbon atoms begins to fall as the radiocarbon decays. The ratio $^{14}C/^{12}C$ in the specimen can be measured and enables the time that has elapsed since the tree was cut down to be calculated. The method has been shown to give consistent results for specimens up to some 40,000 years old, though its accuracy depends upon assumptions concerning the past intensity of the cosmic radiation. The technique was developed by Willard F. Libby (1908-80) and his coworkers in 1946-47.

Carnot's Cycle. A hypothetical scheme for an ideal heat engine. With its aid Carnot proved that the maximum efficiency for the conversion of heat into work depends only on the two

temperatures between which the heat engine works, and not at all on the nature of the substance employed.

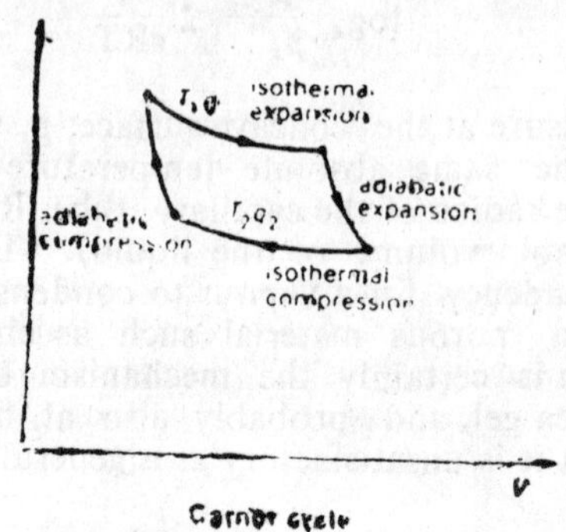

Carnot cycle

It is the most efficient cycle of operations for a reversible heat engine. Published in 1824 by the French physicist N.L.S. Carnot (1746-1832), it consists of four operations on the working substance in the engine :

(*a*) Isothermal expansion at thermodynamic temperature T_1 with heat Q_1 taken in.

(*b*) Adiabatic expansion with a fall of temperature to T_2.

(*c*) Isothermal compression at temperature T_2 with heat Q_2 given out.

(*d*) Adiabatic compression with a rise of temperatare back to T_1.

According to the *Carnot principle,* the efficiency of any reversible heat engine depends only on the temperature range through which it works, rather than the properties of the working substances. In any reversible engine, the efficiency (η) is the ratio of the work down (W) to the heat input (Q_1), *i.e.* $\eta = W/Q_1$. As, according to the first law of thermodynamics, $W = Q_1 - Q_2$, it follows that $\eta = (Q_1 - Q_2)/Q_1$. As the system is reversible. $Q_1/Q_2 = T_1/T_2$, therefore $\eta = (T_1 - T_2)/T_1$. For maximum efficiency T_1 should be as high as possible and T_2 as low as possible.

Carnot's Principle. (Carnot theorem). The efficiency of any heat engine cannot be greater than that of a reversible heat engine operating over the same temperature range. It follows directly from the second law of thermodynamics, and means that all reversible heat engines have the same efficiency, independent of the working substance. If heat is absorbed at temperature T_1 and given out at T_2, then the Carnot efficiency is $(T_1-T_2)/T_1$.

Cascade Liquefier. An appartus which is used for liquefying a gas of low critical temperature. Another gas, already below its critical temperature, is liquified and evporated at a reduced pressure in order to cool the first gas to below its critical temperature. In practice a series of steps is often used, each step enabling the critical temperature of the next gas to be reached.

Cascade Process. Refers to any process that takes place in a number of steps, usually because the single step is too inefficient to produce the desired result. For example, in various uranium-enrichment processes the separation of the desired isotope is only poorly achieved in a single stage ; to achieve better separation the process has to be repeated a number of times, in a series, with the enriched fraction of one stage being fed to the succeeding stage for further enrichment. Another example of cascade process is that operating in a cascade liquefier.

Catalysis. The process of changing the rate of a chemical reaction by use of a catalyst.

Catalyst. A substance which when added to a reaction mixture changes the rate of attainment of equilibrium in the system without itself undergoing a chemical change. Catalysts may be homogenous or heterogeneous, but all comply with the following criteria :

(1) Although the catalyst affects the rate of reaction, it cannot affect the position of equilibrium in a reversible reaction.

(2) In theory, the catalyst can be recovered chemically unchanged at the end of the reaction, although it may be changed physically.

A *positive catalyst* increases the rate of a reaction and a *negative catalyst* reduces it. *Homogeneous catalysts* are those that act in the same phase as the reactants (*i.e.* in gaseous and liquid systems). For example, nitrogen (II) oxide gas will catalyse the reaction between sulpher (IV) oxide and oxygen in the gaseous phase. *Heterogeneous catalysts* act in a different phase from the reactants. For example, finely divided nickel (a solid) will catalyse the hydrogenation of oil (liquid).

The function of a catalyst is to provide a new pathway for which the rate-determining step has a lower activation energy than in the uncatalysed reaction. A catalyst does not change the products in an equilibrium reaction and their concentration is identical to that in the uncatalysed reaction ; *i.e.* the position of the equilibrium remains unchanged. The catalyst simply increases the rate at which equilibrium is attained. In *autocatalysis*, one of the products of the reaction itself acts as a catalyst. In this type of reaction the reaction rate increases with time to a maximum and finally slows down. For example, in the hydrolysis of ethyl acetate, the acetic acid produced catalyses the reaction.

Cataphoresis. See electrophoresis.

Cathetometer. A telescope or microscope fitted with crosswires in the eyepiece and mounted so that it can slide along a graduated scale. Cathetometers are used for accurate measurement of lengths without mechanical contact. The microscope type is often called a *travelling microscope.*

Cathode. The electrode which carries the negative charge in an electrochemical cell. In electrolysis the positively charged ions are discharged at the cathode.

Cathodic Protection. See sacrificial protection.

Cation. A positively charged ion, formed by removal of electrons from atoms or molecules. In electrolysis, cations are attacted to the negatively charged electrode (the cathode). *Compare* anion.

Cell. A device used as a source of electrical energy. Electrochemical cells consist of two electrodes dipping into one or more electrolyte solutions. If the electrodes are in separate solutions, electrical contact between solutions is achieved by means of a salt-bridge, porous disc or ionic conductor. The electrodes may be of the same metal as the electrolyte or of another inert material, *e.g.*, platinum.

An *electrolytic cell* is used for a producing a chemical reaction by passing a current through the electrolyte (*i.e.* by electrolysis). A *voltaic* (or *galvanic*) cell produces an e.m.f. by chemical reactions at each electrode. Electrons are transferred to or from the electrodes, giving each a net charge.

There is a convention for writting cell reaction in voltaic cells. The Daniell cell, for instance, consists of a zinc electrode in a solution of Zn^{2+} ions connected (through a porous pot) to a solution of Cu^{2+} ions in which is placed a copper electrode. The reactions at the electrodes are

$$Zn \rightarrow Zn^{2+} + 2e$$

i.e. oxidation of the zinc to zinc (II), and

$$Cu^{2+} + 2e \rightarrow Cu$$

i.e. reduction of copper (II) to copper. A cell reaction of this type is written :

$$Zn \mid Zn^{2+}(aq) \mid Cu^{2+}(aq) \mid Cu$$

The e.m.f. is the potential of a lead on the right minus the potential of a lead on the left. Copper is positive in this case and the e.m.f. of the cell is stated as +1.10 volts. See also accumulator, Daniell cell, electrolysis, Leclanche cell.

Celsius Scale. Refers to a temperature scale in which the fixed points are the temperatures at standard pressure of ice in equilibrium with water (0°C) and water in equilibrium with steam (100°C). The scale, between these two temperatures, is divided in 100 degree. The degree Celsius (°C) is equal in magnitude to the kelvin. This scale was formerly known as the *centigrade scale* ; the name was officially changed in 1948 to avoid confusion with a hundredth part of a grade. It is named after the Swedish astronomer. Andres Celsius (1701-44), who devised the inverted form of this scale (ice point 100°, steam point 0°) in 1742.

Centi-. Symbol : *c* A prefix denoting 10^{-2}. For example, 1 centimetre (cm) $= 10^{-2}$ metre (m).

Centrigrade Scale. See Celsius scale.

Centrifugal Pump. A device which is commonly used for transporting fluids around a chemical plant. Centrifugal pumps usually have 6-12 blades rotating inside a fixed circular casing. As the blades rotate, the fluid is impelled out of the pump along a pipe. Centrifugal pumps do not produce high pressures but they have the advantage of being relatively cheap because they are simple in design, have no valves, and work at high speeds. In addition they are not damaged if a blockage develops. *Compare* displacement pump.

Centrifuge. A device in which solid or liquid particles of different densities are separated by rotating them in a tube in a horizontal circle. The denser particles tend to move along the length of the tube to a greater radius of rotation, displacing the lighter particles to the other end.

C.G.S. Units. A system of units based on the centimetre, gram, and second. Derived from the metric system, it was badly adapted to use with thermal quantities (based on the inconsistently defined calorie) and with electrical quantities (in which two systems, based respectively on unit permitivity and unit permeability of free space, were used). For scientific purposes c.g.s. units have now been replaced by SI units.

Chain. When two or more atoms form bonds with each other in a molecule, a chain of atoms results. This chain may be a *straight chain*, in which each atom is added to the end of the chain, or it may be a *branched chain*, in which the main chain of atoms has one or more smaller *side chains* branching off it.

Chain Reaction. Refers to a self-sustaining chemical reaction which consists of a series of steps, each of which is initiated by the one before it. An example is the reaction between hydrogen and chlorine :

$$Cl_2 \rightarrow 2Cl.$$

$$H_2 + Cl. \rightarrow HCl + H.$$

$$H. + Cl_2 \rightarrow HCl + Cl.$$

The first stage, chain initiation, has been the dissociation of chlorine molecules into atoms ; this is followed by two chain propagation reactions. Two molecules of hydrogen chloride are produced and the ejected chlorine atom is ready to react with more hydrogen.

Bodenstein and Lind (1906) studied the hydrogen-bromine reaction and concluded that the reaction was not of simple kinetic order. Their rate expression

$$\frac{d[HBr]}{dt}=\frac{k_1[H_2][Br_2]^{\frac{1}{2}}}{k_2+[HBr]/[Br_2]^{\frac{1}{2}}}$$

(where k_1 and k_2 are constants) was later interpreted by means of the chain reaction mechanism :

(*a*) $Br_2 \longrightarrow 2Br.$
(*b*) $Br.+H_2 \longrightarrow HBr+H.$
(*c*) $H.+Br_2 \longrightarrow HBr+Br.$
(*d*) $H.+HBr \longrightarrow H_2+Br.$
(*e*) $2Br. \longrightarrow Br_2$

The first stage (*a*) of the reaction represents the dissociation of a few molecules of bromine into bromine atoms. These atoms then react (*b*) with hydrogen molecules. In step (*c*) bromine atoms are again formed and these can now repeat step (*b*) in the chain propagation. Both steps (*b*) and (*c*) lead to production of HBr, and since bromine atoms are regenerated the process can be repeated. Reaction (*d*) leads to an inhibition in the chain reaction and step (*e*) represents chain termination.

The reaction between hydrogen and chlorine is probably also of this type and many organic free radical reactions (*e.g.* the decomposition of ethanal) proceed via chain mechanisms.

In nuclear chemistry, a fission reaction (see atomic energy) may be initiated by a neutron and may also result in the production of one or more neutrons, which if they reacted in like manner could start a chain reaction. Normally, moderators such as cadmium rods which absorb neutrons are placed in the reactor to control the rate of fission.

Change of Phase (Change of State). A change of matter in one physical phase (solid liquid, or gas) into another. The change is invariably accompanied by the evolution or absorption of energy, even if it takes place at constant temperature (see latent heat).

Charge. Refers to a property of some elementary particles that gives rise to an interaction between them and consequently to the host of material phenomena described as electrical. Charge occurs in nature in two forms, conventionally described as *positive* and *negative* in order to distinguish between the two kinds of interaction between particles. Two particles that have similar charges (both negative or both positive) interact by repelling each other ; two particles that have dissimilar charges (one positive, one negative) interact by attracting each other.

A flow of charged particles, especially a flow of electrons, constitutes an electric current. Charge is measured in coulombs, the charge on an electron being 1.602×10^{-19}. coulombs.

Charles' Law. For a given mass of gas at constant pressure, the volume increases by a constant fraction of the volume at 0°C for each Celsius degree rise in temperature. The constant fraction (a) has almost the same value for all gases—about 1/273—and Charles' law can be written in the form

$$V=V_0(1+\sigma_v\theta)$$

where V is the volume at temperature θ°C and V_0 the volume at 0°C. The constant a_v is the thermal expansivity of the gas. For an ideal gas its value is 1/273.15.

A similar relationship exists for the pressure of a gas heated at constant volume :

$$p=p_0(1+a_p\theta)$$

Here, *ap* is the pressure coefficient. For an ideal gas

$$a_p=a_v$$

although they differ slightly for real gases. It follows from Charles' law that for a gas heated at constant pressure,

$$V/T=K$$

where T is the thermodynamic temperature and K is a constant. Similarly, at constant volume p/T is a constant.

Charles' volume law is sometimes called *Gay-Lussac's law* after its independent discoverer. See *also* absolute temperature, gas laws.

Chemical Cell. See cell.

Chemical Equation. A way of denoting a chemical reaction using the symbols for the participating particles (atoms, molecules, ions etc.) ; for example,

$$xA+yB \rightarrow zC+wD$$

The single arrow is used for an irreversible reaction ; double arrows ($\rightleftharpoons$) are used for reversible reactions. When reactions involve different phases it is usual to put the phase in brackets after the symbol (s=soid ; l=liquid ; g=gas ; aq=aqueous). The numbers x, y, z and w. showing the relative numbers of molecules reacting, are called the *stoichiometric coefficients.* The sum of the coefficients of the reactants minus the sum of the coefficients of the products ($x+y-z-w$ in the example) is the *stoichiometric sum.* If this is zero the equation is balanced. Sometimes a generalized chemical equation is considered

$$v_1A$$

In this case the reaction can be written $\Sigma v_i A_i=0$, where the convention is that stoichiometric coefficients are positive for reactants and negative for products. The stoichiometric sum is Σv_i.

Chemical Equilibrium. A reversible chemical reaction in which the concentrations of reactants and products are not changing with time because the system is in thermodynamic equilibrium. For example, the reversible reaction

$$3H_2+N_2 \rightleftharpoons 2NH_3$$

is in chemical equilibrium when the rate of the *forward reaction*

$$3H_2+N_2 \rightarrow 2NH_3$$

is equal to the rate of the *back reaction*

$$2NH_3 \rightarrow 3H_2+N_2$$

See *also* equilibrium constant.

Chemical Potential. Symbol : μ For a given component in a mixture, the coefficient $\delta G/\delta n$, where G is the Gibbs free energy and n the amount of substance of the component. The chemical potential is the change in Gibbs free energy with respect to change in amount of the component, with pressure, temperature, and amounts of other components being constant. Components are in equilibrium if their chemical potentials are equal.

Chemical potential is a valuable function in the theory of heterogeneous equilibria, since it can readily be proved that, if such a system is not initially at equalibrium, substances will tend to pass from one phase to another (*e.g.*, solid dissolving or crystallizing) until a uniform chemical potential is attained for each substance in every phase.

Chemical Shift. See nuclear magnetic resonance.

Chemiluminescence. A number of chemical reactions are accompanied by emission of light, *e.g.*, the oxidation of yellow phosphorus. The process has been the reverse of the ordinary photochemical one, in which reactions are caused to take place by exposing the reactants to light, and is termed chemiluminescence. The light emitted by the fire-fly or glow-worm, and indeed ordinary luminous combustion, are examples of this very common phenomenon.

Chemiosmotic Hypothesis. Refers to the hypothesis that the function of the electron transport chain is to translocate protons and thereby create a pH difference and a membrane potential. This is termed as the proton motive force and is used by the proton pump of the mitochondrial ATP-ase operating in reverse to generate ATP. Evidence has gradually been gathered to suggest that this may broadly be true.

Chemisorption. In many cases the adsorption of a substance at a surface involves the formation of chemical bonds between the adsorbate, the species undergoing adsorption and the adsorbent, the surface at which adsorption is occurring. Such a process is termed chemisorption. It is of importance in heterogeneous catalysis.

Chemistry. The study of the elements and the compounds they form. Chemistry is mainly concerned with effects that depend on the outer electrons in atoms. See biochemistry ; geochemistry ; inorganic chemistry ; organic chemistry ; physical chemistry.

Circular Dichroism. In the Cotton effect distinct changes in the polarization angle are seen as a function of wavelength. This is due to differences in the specific extinction coefficients for left- and right-handed polarized light. As a linearly-polarized light wave passes through a substance the differing absorptions cause an elliptically polarized wave to be produced. This phenomenon is known as circular dichroism. The magnitude of the effect is expressed by the equation

$$\phi=\pi/\lambda(\eta_1-\eta_r)$$

where ϕ is the ellipticity (in radians) of the emerging beam and η_1 and η_r are the absorption indices of the left- and right-handed circularly-polarized light respectively. When the ellipticity is plotted as a function of wavelength, a curve is given with a maximum corresponding to the wavelength of zero angle in the optical rotatory dispersion curve. Optical isomers give circular dichroism curves which are identical except that in one case the effect is positive, *i.e.*, ϕ is positive throughout whereas for the other isomer the effect is negative.

Clapeyron-Clausius Equation. A thermodynamic equation applying to any two-phase equilibrium for a pure substance. The equation states :

$$\frac{dP}{dT}=\frac{\Delta H_v}{T\Delta V}$$

where P is the pressure, T the absolute temperature, ΔH_v the molar latent heat of the vaporization and ΔV the corresponding change in molar volume. If one phase is the vapour, the equation reduces to

$$\ln P=C-\Delta H_v/RT$$

where C is a constant for the particular substance and R is the gas constant.

Clark Cell. A type of cell formerly used as a standard source of e.m.f. It consists of a mercury cathode coated with mercury sulphate, and a zine anode. The electrolyte is zinc sulphate solution. The e.m.f. produced is 1.4345 volts at 15°C. The Clark cell has been superseded as a standard by the Weston cadmium cell.

Claude Process. A process similar to the Linde process for the liquefaction of air, except that additional cooling is produced by allowing the expanding gas to do external work.

Clausius-Mosotti Law. The molecular polarization (P) of a substance whose molecular weight is M and whose density is *d* and dielectric constant D is given by this law which states :

$$P=\frac{(D-1)}{(D+2)} \cdot \frac{M}{d}$$

Cleavage. The splitting of a crystal along planes of atoms, to form smooth surfaces.

Close-Packed Structures. The majority of metals, alloys and ionic inorganic compounds, and also many compounds which are appreciably covalent, have structures which can be described in terms of close-packing of the largest species, atoms or ions. In each layer of a true close-packed structure each atom has fix neighbours in an hexagonal array. A similar second layer lies with its atoms above the holes of the first layer. The atoms of the third layer can be directly above the atoms of the first layer (hexagonal close-packing, hcp) or above holes in both first and second layers (cubic close-packing, ccp). In the

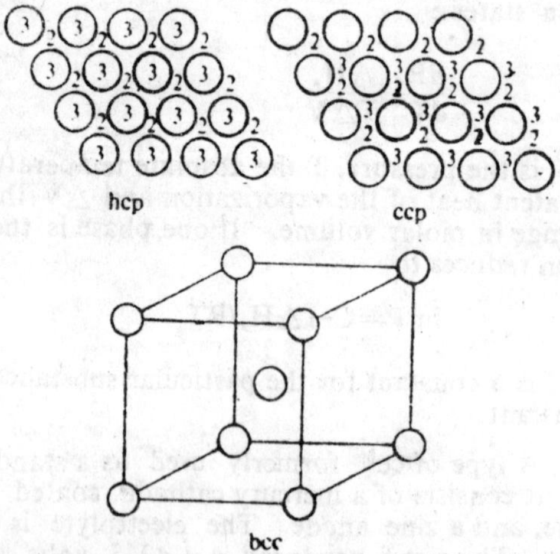

latter case the fourth layer lies over the first. More complex sequences of layers are known. The body-centred cubic structure, bcc, does not make quite such full use of space as hcc or ccp. Many structures are described in terms of smaller cations occupying appropriate holes (octahedral △ or tetrahedral T) between. For each close-packed atom there is one octahedral hole and two tetrahedral holes (in the diagrams the number *n* designates an atom in the *n*th layer, the first layer is designated O).

Close Packing. The arrangement of particles (usually atoms) in crystalline solids in which each particle has 12 nearest neighbours : six in the same layer (or plane) as itself and three each in the layer above and below. This arrangement provides the most economical use of space. The two principal types are cubic close packing and hexagonal close packing.

Clusius Column. A device for separating isotopes by thermal diffusion. One form consists of a vertical column some 30 metres high with a heated electric wire running along its axis. The lighter isotopes in a gaseous mixture of isotopes diffuse faster than the heavier isotopes. Heated by the axial wire, and assisted by natural convection, the lighter atoms are carried to the top of the column, where a fraction rich in lighter isotopes can be removed for further enrichment.

Coacervation. Under certain conditions lyophilic (hydrated colloid) sols may be made to separate into two immiscible liquid phases, each of which has a different concentration of the dispersed phase. This process is termed coacervation and the two liquids are called coacervates. An example is the coacervation of protein sols by alcohol at 50°C.

Coagel. A gelatinous precipitate ; *e.g.*, rigid silica gel.

Coagulation. The process in which colloidal particles come together to form larger masses. Coagulation can be brought about by adding ions to neutralize the charges stabilizing the colloid. Ions with a high charge are particularly effective (*e.g.* alum, containing Al^{3+}, is used in styptics to coagulate blood). Another example of ionic coagulation is in the formation of river deltas, which occurs when colloidal silt particles in rivers are coagulated by ions in sea water. Heating is another way of coagulating certain colloids (*e.g.* boiling an egg coagulates the albumin).

Coherent Units. A system or sub-set of units (*e.g.* SI units) in which the derived units are obtained by multiplying or dividing together base units, with no numerical factor involved.

Colligative Properties. A group of properties of solutions that depends on the number of particles present, rather than the nature of the particles. Colligative properties include :

(1) The lowering of vapour pressure.

(2) The elevation of boiling point.

(3) The lowering of freezing point.

(4) Osmotic pressure.

Colligative properties are all based upon empirical observation. The explanation of these closely related phenomena deprnds on intermolecular forces and the kinetic behaviour of the particles, which is qualitatively similar to those used in deriving the kinetic theory of gases.

Collimator. An arrangement for producing a parallel beam of radiation for use in a spectrometer or other instument. A system of lenses and slits is utilized.

Colloid. Refers to a heterogeneous system in which the interfaces between phases, though not visibly apparent, are important factors in determining the system properties. The three important attributes of colloids are :

(1) They contain particles, commonly made up of large numbers of molecules, forming the distinctive unit or *disperse phase.*

(2) The particles are distributed in a continuous medium (the *continuous phase*).

(3) There is a stabilizing agent, which has an affinity for both the particle and the medium ; in many cases the stabilizer is a polar group.

Particles in the disperse phase typically have diameters in the range 10^{-6}-10^{-4} mm. Milk, rubber, and emulsion paints are typical examples of colloids. See *also* sol.

Sols are dispersions of small solid particles in a liquid. The particles may be macromolecules or may be clusters of small molecules. *Lyophobic sols* are those in which there is no affinity between the dispersed phase and the liquid. An example is silver cloride dispersed in water. In such colloids the solid particles have a surface charge, which tends to stop them coming together. Lyophobic sols are inherently unstable and in time the particles aggregate and form a precipitate. *Lyophilic sols*, on the other hand, are more like true solutions in which the solute molecules are large and have an affinity for the solvent. Starch in water is an example of such a system. *Association colloids* are systems in which the dispersed phase consists of clusters of molecules that have lyophobic and lyophilic parts. Soap in water is an association colloid (see micelle).

Emulsions are colloidal systems in which the dispersed and continuous phases are both liquids, *e.g.* oil-in-water or water-in-oil. Such systems require an emulsifying agent to stabilize the dispersed particles.

Gels are colloids in which both dispersed and continuous phases have a three-dimensional network throughout the material, so that it forms a jelly-like mass. Gelatin is a common example. One component may sometimes be removed (*e.g.* by heating) to leave a rigid gel (*e.g.* silica gel).

Other types of colloid include *aerosols* (dispersions of liquid or solid particles in a gas, as in a mist or smoke) and foams (dispersions of gases in liquids or solids).

Colloidal Electrolyte. Many colloidal materials react as if they consisted of numbers of large polyvalent ions in equilibrium with the corresponding number of simple ions of opposite charge. Thus, although the sodium salts of the lower fatty acids behave as normal weak electrolytes, with higher membranes of the series, association begins to occur and colloidal character is assumed. The anion is no longer a discrete fatty acid anion, but consists of an aggregate of fatty acid anions together with undissociated salt molecules ; it is a giant polyvalent ion, dissociating at the surface and compensated by an equal number of sodium ions in close proximity to the giant ion. Soaps and many dyestuffs may be classed as colloidal electrolytes. When electrolysed the rate of movement of the giant ions towards the anode is similar to that of simple ions and transport number determinations can be carried out normally.

Colloid Mills. Devices for producing colloidal suspensions or emulsions where the particle and droplet sizes are less than one micron. By producing very high shearing forces within the fluid they perform size reduction and dispersion at the same time. These forces may be obtained in various ways.

Colloid mills are used in the paint industry, and in the pharmaceutical and food-processing industries.

Colorimetric Analysis. Quantitative analysis in which the concentration of a coloured solute is measured by the intensity of the colour. The test solution can be compared against standard solutions.

Combustion. A chemical reaction in which a substance reacts rapidly with oxygen with the production of heat and light. Such reactions are often free-radical chain reactions, which can usually be summarized as the oxidation of carbon to form its oxides and the oxidation of hydrogen to form water. See *also* flame.

Common-Ion Effect. In a solution of a weak electrolyte, *e.g.* ethanoic acid, the concentration of ions is governed by the equilibrium

$$HAc \rightleftharpoons H^+ + Ac^-$$

Addition of excess H^+ ions to this solution will cause the equilibrium to move towards undissociated acid thereby decreasing the concentration of Ac^-. This effect is known as the common-ion effect and is of considerable practical importance. Thus, *e.g.* in the precipitation of metal ions as insoluble sulphides, the concentration of S^{2-} in aqueous solution is controlled by the equilibrium

$$H_2S(aq) \rightleftharpoons 2H^+(aq) + S^{2-}(aq)$$

Addition of acid will reduce the concentration of S^{2-}, whilst in alkaline solution the concentration of S^{2-} will increase. Since, in order for precipitation to occur, the solubility product of the sulphide must be exceeded, *i.e.*,

$$K_{sp} = [M^{x+}][S^{2-}]^{x/2}$$

the actual precipitation can be controlled by varying the pH of the solution.

Component. A distinct chemical species in a mixture. If there are no reactions taking place, the number of components is the number of separate chemical species. A mixture of water and ethanol, for instance, has two components (but is a single phase). A mixture of ice and water has two phases but one component (H_2O). If an equilibrium reaction occurs, the number of components is taken to be the number of chemical species minus the number of reactions. Thus, in

$$H_2+I_2 \rightleftharpoons 2HI$$

there are two components. See *also* phase rule.

Concentrated. Containing a high proportion (*e.g.*, of solute in a solution).

Concentration. The amount of substance per unit volume in a solution. The unit is the mol dm^{-3}. *Mass concentration* is mass of solute per unit volume. *Molal concentration* is amount of substance per kilogram of solute.

Concentration Cell. The potential of a piece of metal immersed in a solution containing its ions (see electrode potential) depends on the concentration of the ions. Thus a cell may be set up which derives its electromotive force from the difference in concentration of solutions of the same electrolyte surrounding the two electrodes. Any cell which depends on this principle is called a concentration cell.

Conchoidal Fracture. A surface resulting from the fracture of a solid which shows no regular crystal faces (cleavage planes). The fracture is generally made up of shell-like curved surfaces and is characteristic of amorphous glassy materials.

Condensation. The conversion of a gas or vapour into a liquid or solid by cooling.

Conductiometric Titration. The term used for a titration in which measurement of the electrical conductance is made continuously throughout the addition of the titrant and well beyond the equivalence point. This is in place of traditional end-point determination by indicators. The operation is carried out in a conductance cell, which is part of a resistance bridge circuit. The method depends on the fact that ions have different ionic mobilities, H^+ and OH^- having particularly high values. The method is especially useful for weak acid-strong base and strong acid-weak base titrations for which colour-change titrations are unreliable.

Conductivity. The property of a substance by virtue of which the substance allows the passage of an electric current is termed the conductivity. The reciprocal of the resistance of a circuit is called the conductivity. The reciprocal of the specific resistance is the specific conductivity or conductance of a substance.

The conductivity of a solution containing 1 gram equivalent of solute when measured between two large parallel electrodes at a distance of 1 cm apart is called the equivalent conductivity Δ.

$$\Delta = \frac{1000\ K}{c}$$

where K=specific conductance and c=concentration in g equiv. per litre.

Conductivity, Solids. Solids may conveniently be classified in terms of their electrical conductivity into electronic conductors, semiconductors and insulators. Metals are typical conducting solids, their conductivity increasing with decreasing temperature. Semiconductors may be classified as either intrinsic semiconductors, *e.g.*, pure germanium or silicon and a number of transition metal oxides, or as impurity semiconductors, *e.g.* Al or P doped Si and Ge. Semiconductors may be differentiated from conducting solids in that the conductivity of the former increases markedly with increasing temperature. Refractory oxides, *e.g.*, alumina, silica, magnesium oxide are typical insulators. Impurity doped semiconductors find extensive use in electronic microcircuits as transistors, etc. Ionic conductors conduct by movement of ions through the solid.

Conjugate Base. In Bronsted-Lowry acid-base theory an acid is considered to dissociate into a proton and the conjugate base of the acid. Thus the conjugate base of an acid is its anion.

Conjugate Solutions. Solutions of two substances in one another (*e.g.* phenol in water and water in phenol) which are in equilibrium at a particular temperature.

Conservation of Energy, Law of. Enunciated by Helmholtz in 1847, this law states that in all processes occurring in an isolated system the energy of the system remains constant. The law does of course permit energy to be converted from one form to another.

Consolute Temperature. The temperature at which two partially miscible liquids become fully miscible as the temperature is increased.

Constant-Boiling Mixture. A general observation for most liquids has been that the vapour phase above a liquid is richer in the more volatile component (deviation from Raoult's law). Consequently most liquid mixtures gives a regular increase in the boiling point as the liquid is progressively distilled. The boiling point-composition curve reveals that distillation of the liquid of composition L_1 gives a vapour richer in A and represented by composition V_1. Further distillation leads to the liquid composition moving towards B. For certain mixtures in which there are strong inter-molecular attractions the boiling point-composition curves show minima or maxima. Fractional distillation of the former leads to initial changes in the vapour until a distillate of composition L_2 is reached at which point a constant boiling mixture or azeotrope distils over. Further attempts to fractionate the distillate do not lead to a change in composition. An example of an azeotropic mixture of minimum boiling point is water (b.p 100°C) and ethanol (b.p. 78.3°C), the azeotrope being 4.4% water and boiling at 78.1°C.

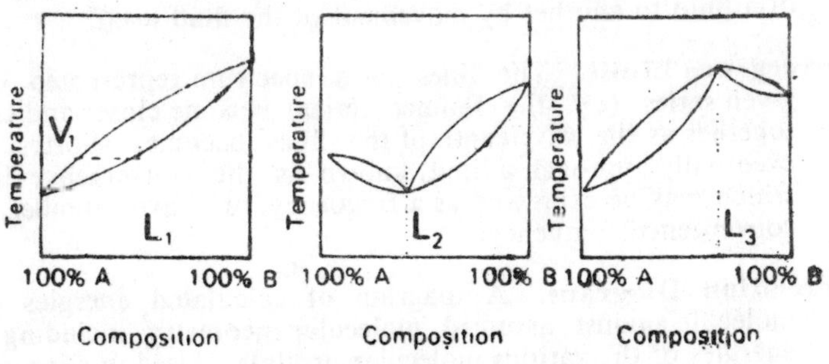

Constant – boiling mixtures

Mixtures that display a maximum in the boiling point-composition curve can lead to initial separation of pure A on fractionation but as the composition of the liquid moves towards B and reaches the maximum, a constant boiling mixture L is reached that will distil over unchanged. An example of an azeotropic mixture of maximum boiling point is water (b.p. 100°C) and hydrogen chloride (b.p. -80°C), the azeotrope being 80% water and boiling at 108.6°C. See *also* azeotrope.

Continuity of State. As the temperature gets increased towards the critical point, the properties of liquids and vapour become increasingly similar, until at the critical temperature they become identical. Although the change from liquid to vapour, or vice-versa, is normally discontinuous, a gradual transition with continuity of state is possible. For example, if carbon dioxide at one atmosphere pressure and 293 K is isothermally compressed to 100 atmospheres, the appearance of liquid and disappearance of vapour is sharply defined. However, if the carbon dioxide is first heated to 313 K, isothermally compressed to 100 atmospheres then cooled to 293 K the same liquid state as before is reached without any discontinuity.

Continuous Phase. See colloid.

Continuous Spectrum. A spectrum composed of a continuous range of emitted or absorbed radiation. Continuous spectra are produced in the infrared and visible regions by hot solids. See *also* spectrum.

Convection. A process by which heat is transferred from one part of a fluid to another by movement of the fluid itself.

Convergence Limit. The lines in a spectrum represented by a given series (*e.g.* the Balmer series) become closer and closer together as the wavelength of the lines becomes shorter, and eventually approach a limit, known as the convergence limit, which may be expressed as a frequency, or wave number, the convergence frequency.

Correlation Diagram. A diagram of calculated energies of a molecule against assumed molecular geometry including the energies of the various molecular orbitals. Used in approaches to molecular geometry.

Cotton Effect. Within an absorption band, there is anomalous rotatory dispersion. In the figure, the broken line represents an absorption band and the full line the rotatory dispersion curve. As the wavelength decreases the rotation angle increases, passing through a maximum and then decreases, passing through zero angle at the wavelength corresponding to the maximum of absorption. On proceeding to lower wavelengths, the angle falls till it passes through a minimum and then rises again. This phenomenon is known as the cotton effect. This effect is general for coloured compounds as well as for colourless substances with bands in the ultra-violet. See optical rotatory dispersion, circular dichroism.

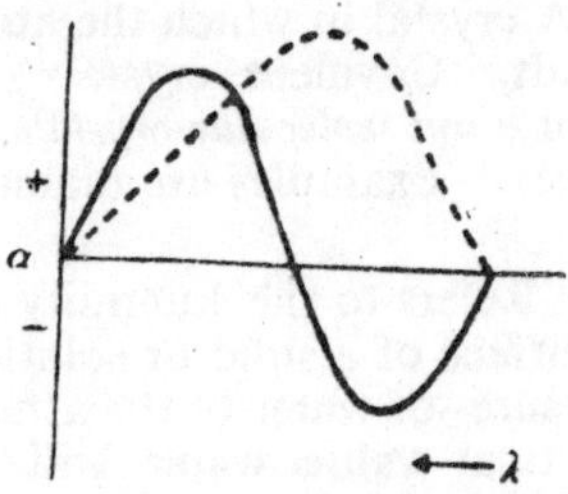

Cottrell Precipitator. A common name for the single-stage electrostatic precipitator. See electrostatic precipitators.

Coulomb. Symbol :C The SI unit of electric charge, equal to the charge transported by an electric current of one ampere flowing for one second. 1 C=1 As.

Coulometer An instrument or device which is used for measuring the amount of charge (number of coulombs) passing in an electrical circuit. Thus a silver coulometer consists of 2 platinum electrodes dipping into a solution of silver nitrate. For each coulomb of charge passed 1.118×10^{-3}g of silver is deposited at the cathode. Thus, from the increase in weight of the cathode the charge passing can be calculated.

Coulometry. Analytical technique which is involving the measurement of quantities of electricity. Constant current coulometry can be used for coulometric titrations, *e.g.* Br_2 can be prepared electrically and used for oxidation (*e.g.*, of olefines) ***in situ*** and the end point detected spectrophotometrically ; Ag^+ can be generated and used to titrate Cl^-. In controlled (constant) potential coulometry one species may be oxidized or reduced and thus determined at a specific potential.

Counter-Ions. The charge on the surface of a colloidal particle is compensated for by an equal and opposite ionic charge in the liquid in immediate contact with it. The system as a whole is electrically neutral. These compensating ions, which occupy a diffuse layer around the particle, are termed counter-ions.

Also used to describe the oppositely charged ion balancing charge, *e.g.*, chloride is the counter-ion to Na^+ in sodium chloride.

Covalent Crystal. A crystal in which the atoms are held together by covalent bonds. Covalent crystals are sometimes called *macromolecular* or *giant-molecular crystals.* They are hard high-melting substances. Examples are diamond and boron nitride.

Critical Humidity. Refers to the humidity at which the vapour pressure at the surface of a solid or solution becomes equal to the partial pressure of water in the atmosphere. At humidities above the critical value water will tend to be absorbed; below the critical value moisture will be lost to the atmosphere.

Critical Phenomena. Above the critical temperature it is impossible to liquefy a gas. The minimum pressure required to cause liquefaction at the critical temperature is the critical pressure. The critical volume is the volume occupied by one mole of the gas at its critical temperature and pressure.

Critical Point. The conditions of temperature and pressure under which a liquid being heated in a closed vessel becomes indistinguishable from the gas or vapour phase. At temperatures below the critical temperature (T_c) the substance can be liquefied by applying pressure; at temperatures above T_c this is not possible. For each substance there is one critical point; for example, for carbon dioxide it is at 31.1°C and 73.0 atmospheres.

Critical Pressure. The lowest pressure needed to bring about liquefaction of a gas at its critical temperature.

Critical Solution Temperature. The temperature at which the mutual solubility of two phases becomes equal. For two liquids, *e.g.*, phenol and water, the solubility of each in the other increases until at the critical solution temperature only one phase exists.

Critical State. The state of a fluid in which the liquid and gas phases both have the same density. The fluid is then at its critical temperature, critical pressure, and critical volume.

Critical Temperature. The temperature below which a gas can be liquefied by applying pressure and above which no amount of pressure is sufficient to bring about liquefaction. Some gases have critical temperatures above room temperature (*e.g.*, carbon dioxide 31.1°C and chlorine 144°C) and have been known in the liquid state for many years. Liquefaction proved much more difficult for those gases (*e.g.*, oxygen −118°C and nitrogen—146°C) that have very low critical temperatures.

Critical Volume. The volume of a fixed mass of a fluid in its critical state ; *i.e.*, when it is at its critical temperature and critical pressure. The *critical specific volume* is its volume per unit mass in this state : in the past this has often been called the critical volume.

Cross Section. A number denoting the effective area of a nucleus for a scattering or absorption process. See barn.

Cryohydrate. A eutectic mixture of ice and some other substance (*e.g.*, an ionic salt) obtained by freezing a solution.

Cryoscopic Constant. See depression of freezing point.

Cryostat. A vessel enabling a sample to be maintained at a very low temperature. The Dewar flask is the most satisfactory vessel for controlling heat leaking in by radiation, conduction, or convection. Cryostats usually consist of two or more Dewar flasks nesting in each other. For example, a liquid nitrogen bath is often used to cool a Dewar flask containing a liquid helium bath.

Crystal. A solid substance that has a definite geometric shape. A crystal has fixed angles between its faces, which have distinct edges. The crystal will sparkle if the faces are able to reflect light. The constant angles are caused by the regular arrangements of particles (atoms, ions, or molecules) in the crystal. If broken, a large crystal will form smaller crystals.

In crystals, the atoms, ions, or molecules of the substance form a distinct regular array in the solid state. The faces and their angles bear a definite relationship to the arrangement of these particles.

Crystal Drawing. Crystals are drawn in a conventional projection and in a conventional position. For the cubic system the x-axis does not point to the eye of the observer, but is rotated clockwise through an angle of 18°26′, the tangent of which is $\frac{1}{3}$; the y-axis is not drawn at right angles to the vertical c axis because the eye is elevated at an angle of 9°28′, the tangent of which is $\frac{1}{6}$. The axes are therefore drawn as shown (*next column*).

Crystal Habit. The shape of a crystal. The habit depends on the way in which the crystal has grown; *i.e.*, the relative rates of development of different faces.

Crystalline State. The form of the solid state with an ordered arrangement of atoms, etc. in the solid. The pattern is characteristic of the solid and generally gives rise to characteristic bounding faces of the crystal.

Crystal Lattice. The regular pattern of atoms, ions, or molecules in a crystalline substance. A crystal lattice can be regarded as produced by repeated translations of a *unit cell* of the lattice. See also crystal system.

Crystalline. Having the regular internal arrangement of atoms, ions, or molecules characteristic of crystals. Crystalline materials need not necessarily exist as crystals; all metals, for example, are crystalline although they are not usually seen as regular geometric crystals.

Crystallite. A small crystal that has the potential to grow larger. It is often used in mineralogy to describe specimens that contain accumulations of many minute crystals of unknown chemical composition and crystal structure.

Crystallization. The process of forming crystals from a liquid or gas.

Crystallographic Axes. The (arbitrary) axes defining the repeat unit (unit cell) in a crystal. See crystal structure.

Crystal Nucleus. A minute crystal which serves as a centre of formation for larger crystals which would not otherwise be formed. *e.g.*, crystallization of a supersaturated solution may be induced by formation of a crystal nucleus through mechanical shock, inoculation by dust, etc.

Crystals, Liquid. See liquid crystals.

Crystal Structure. The arrangement of atoms and molecules in a crystal, as revealed by X-ray and electron diffraction methods.

The shape of the unit cell, and hence the constants required to define it, depends on the symmetry of the crystal, and the following table summarizes the data required in the various systems :

Triclinic $a, b, c, \alpha, \beta, \gamma$ (no planes or axes of symmetry).

Monoclinic a, b, c, β, $(\alpha=\gamma=90°)$—one 2-fold axis and/or one plane of symmetry.

Orthorhombic a, b, c, $(\alpha=\beta=\gamma=90°)$—one 2-fold axis at the intersection of two planes or perpendicular to two other 2-fold axes.

Trigonal a $(=b)$ c, $(\gamma=120°, \alpha=\beta=90°)$—one 6-fold axis perpendicular to other axes of symmetry.

Hexagonal a $(=b)$ c, $(\delta=120°, \alpha=\beta=90°)$—one 6-fold axis perpendicular to other axes of symmetry.

Tetragonal a $(=b)$, c, $(\alpha=\beta=\gamma=90°)$—one 4-fold axis perpendicular to other axes of symmetry.

Cubic a $(=b, =c)$ $(\alpha=\beta=\gamma=90°)$—four 3-fold axes at the tetrahedral angles to one another.

The crystal structure determines not only the arrangement of atoms in the lattice but also the external form of the crystal.

Crystallography. The study of crystal form and structure. See also X-ray crystallography.

Crystal Symmetry. A term describing the regularities in the position and arrangement of faces and edges of a crystal and also of the atoms within the crystal. Such regularities are defined in terms of planes of symmetry, axes of symmetry and centres of symmetry.

Crystal Systems. Crystals are placed in one of the seven crystal systems : cubic, tetragonal, hexagonal, trigonal, orthorhombic, monoclinic and triclinic; according to their symmetry. For the characteristics of each of these systems see crystal structure and the separate headings (*e.g.*, cubic system).

If the unit cell is a parallelopiped with lengths *a*, *b*, and *c* and the angles between these edges are α (between *b* and *c*), β (between *a* and *c*), and γ (between *a* and *b*), then the classification is:

cubic : $a=b=c$; $\alpha=\beta=\gamma=90°$
tetragonal : $a=b=\neq c$; $\alpha=\beta=\gamma=90°$
orthorhombic : $a\neq b\neq c$; $\alpha=\beta=\gamma=90°$
hexagonal : $a=b\neq c$; $\alpha=\beta=90°$; $\gamma=120°$
trigonal : $a=b\neq c$; $\alpha=\beta=\gamma\neq 90°$
monoclinic : $a\neq b\neq c$; $\alpha=\gamma=90°\neq\beta$
triclinic : $a\neq b\neq c$; $\alpha\neq\beta\neq\gamma$

The orthorhombic system is also called the *rhombic* system.

Cubic. Denoting a crystal in which the unit cell is a cube. In a *simple cubic* crystal the particles are arranged at the corners of a cube. See also body-centred cubic crystal, face-centred cubic crystal, crystal system.

Cubic Close Packed. See face-centred cubic crystal.

Cubic Crystal. A crystal in which the unit cell is a cube (see crystal system). There are three possible packings for cubic crystals : *simple cvbic, face-centred cubic,* and *bodycentred cubic.* See illustration.

Cubic System. The crystal system with three equal crystallographic axes at right angles, *e.g.*, NaCl.

Cubic Close-Packing, CCP. See close-packed structure.

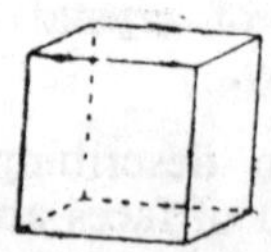

Cubic Co-ordination. Co-ordination by eight ligands at the corners of a cube as in CsCl. A cube has 6 square faces. Compare square-antiprism and dodecahedron.

Curie. The standard unit of radioactivity. 1 Cl=3.7×10^{10} disintegrations per second ; the same number of disintegrations as produced by 1 g of radium.

Curie Point (Curie Temperature). The temperature at which a ferromagnetic substance loses its ferromagnetism and becomes only paramagnetic. For iron the Curie point is 760°C and for nickel 356°C.

Curie-Weiss Law. The molar magnetic susceptibility of a substance may be represented as :

$$\chi = C/T,$$

where C is the Curic constant, and χ contains both diamagnetic and paramagnetic contributions. Not all substances conform to this equation ; a more general relationship, the Curie-Weiss law, states :

$$\chi = \frac{C}{T-\theta}$$

θ is the Weiss constant and is characteristic of the particular substance being considered.

Cyclic Process. Refers to a system which has undergone a series of changes and has returned to its original state has completed a cycle. The whole series of changes is a cyclic process. See *e.g.*, Born-Haber cycle.

Cyclic Voltammetry. Analytical method by which a species is successively oxidized and reduced by application of a varying potential. The current is plotted against voltage.

Cyclone Separator. A unit which is used for removing suspended solids from air or other gases. It is usually in the form of a vertical cylinder with a flat top and a conical lower end Gas is introduced at high velocity through a tangential inlet near the top, and the swirling motion that results causes the particles to be thrown out against the walls by centrifugal force. The motion of the gas is such that the solids are carried into the conical end from where they may be discharged ; the clean gas leaves through an axial outlet at the other end. Because the movement of particles into the conical end is not dependent on gravity, cyclones may be, and frequently are, mounted horizontally.

The cyclone separator has been the most widely used type of dust collection plant, being simple and cheap. It is very effective provided the proportion of particles below 10μ is not too large.

Cyclotron. An apparatus which is used to accelerate charged atomic particles, *e.g.*, protons, by passing them repeatedly through the same electric field. By this method particles of very great velocity and hence of great energy can be produced.

D

Daniell Cell. A type of primary voltaic cell with a copper positive electrode and a negative electrode of a zinc amalgam. The zinc amalgam electrode is placed in an electrolyte of dilute sulphuric acid or zinc sulphate solution in a porous pot, which stands in a solution of copper sulphate in which the copper electrode is immersed. While the reaction takes place ions move through the porous pot, but when it is not in use the cell should be dismantled to prevent the diffusion of one electrolyte into the other. The e.m.f. of the cell is 1.08 volts with sulphuric acid and 1.10 volts with zinc sulphate. It was invented in 1836 by the British chemist John Daniell (1790-1845).

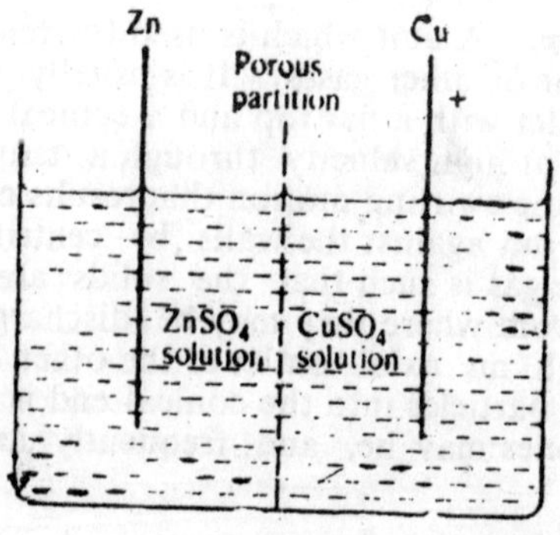

The e.m.f. of the Daniell cell is virtually independent of temperature.

The positive electrode is copper immersed in copper (II) sulphate solution. The negative electrode is zinc-mercury amalgam in either dilute sulphuric acid or zinc sulphate solution. The porous pot prevents mixing of the electrolytes, but allows ions to pass. With sulphuric acid the e.m.f. is about 1.08 volts ; with zinc sulphate it is about 1.10 volts.

At the copper electrode copper ions in solution gain electrons from the metal and are deposited as copper atoms :

$$Cu^{2+} + 2e \rightarrow Cu$$

The copper electrode thus gains a positive charge. At the zinc electrode, zinc atoms from the electrode lose electrons and dissolve into solution as zinc ions, leaving a net negative charge on the electrode

$$Zn \rightarrow 2e + Zn^{2+}$$

Dalton's Law. (of partial pressures). The pressure of a mixture of gases is the sum of the partial pressures of each individual constituent. (The *partial pressure* of a gas in a mixture is the pressure that it would exert if it alone were present in the container.) Dalton's law is strictly true only for ideal gases.

Deactivating Collision. Refers to an intermolecular collision between an activated and a normal molecule, which results in the deactivation of the former without it undergoing reaction. See deactivation.

Deactivation. Refers to the process whereby the chemical reactivity of a substance gets diminished or entirely removed. For example, the catalytic activity of metals like platinum or nickel for various hydrocarbon reactions may be considerably reduced or completely removed by the presence of trace amounts of catalyst poisons such as sulphur or mercury which cause deactivation of the catalyst.

Alternatively, in a photochemical reaction a molecule absorbs light. This causes it possessing more energy than an ordinary molecule and, con sequently, more reactivity. However, collision of this activated molecule with another molecule may make the former to lose its energy without reacting. The activated molecule is said to have undergone a deactivating collision and has itself become deactivated.

Deactivators. Additives which are used in gasoline to prevent the catalysis of gum formation by metals (*e.g.* Cu). Deactivators act by passivation of the metal surface, preventing solution in gasoline or by deactivation of the metal compounds already in the fuel. Commercial deactivators contain amine or diamine groups which combine with the metal to form non-soluble, non-catalytic compounds.

Darcy. Refers to a unit of permeability for flow or fluid through a porous bed which is moving under the influence of an applied pressure gradient. The constant k in the following equation (Darcy's Law) may be expressed in darcys when Q is the volume rate of flow (cm^3/s) ; A is the cross-sectional area (cm^2) ; μ is viscosity of fluid (cP) and dP/dL is the pressure gradient at right angles to the cross-section (atm/cm).

$$Q = \frac{Ak}{\mu} \cdot \frac{dP}{dL}$$

De Broglie Equation. In 1924, de Broglie suggested that the wavelength of electrons (λ) is given by the equation :

$$\lambda = (h/p) = (h/mv)$$

where h=Planck's constant, m=mass of electron, v=velocity and p=momentum of the electron. Wavelengths calculated from this equation are often called de Broglie wavelengths.

De Broglie Wavelength. The wavelength of the wave associated with a moving particle. The wavelength (λ) is given by $\lambda = h/mv$, where h is the Planck constant, m is the mass of the particle, and v its velocity. The *de Broglie wave* was first suggested by the French physicist Louis de Broglie (1892-) in 1924 on the grounds that electromagnetic waves can be treated as particles (photons) and one could therefore expect particles to behave in some circumstances like waves. The subsequent observation of electron diffraction substantiated this argument and the de Broglie wave became the basis of wave mechanics.

Debye. **Symbol** : D. A unit of electric dipole moment equal to 3.33564×10^{-30} coulomb metre. It is used in expressing the dipole moments of molecules.

Debye-Huckel Theory. The activity coefficient of an electrolyte depends markedly upon concentration. In dilute solutions due to the Coulombic forces of attraction and repulsion the ions tend to surround themselves with an 'atmosphere' of oppositely charged ions. Debye and Huckel showed that it was possible to explain the abnormal activity coefficients at least for very dilute solutions of electrolytes. For dilute solutions, the Debye-Huckel equation by calculation based on these Coulombic interactions is

$$\ln \gamma_i = \frac{-\epsilon^3 z_i^2}{(\epsilon kT)^{\frac{1}{2}}} \sqrt{\frac{2\pi LI}{1000}}$$

where γ_i is activity coefficient of the ion i ; z_i is the ionic charge ; e is the electronic charge ; ϵ, the di-electric constant of the solution ; L, Avagadro's number and I is the ionic strength of the solution.

Deca-. Symbol da. A prefix which is used in the metric system to denote ten times. For example, 10 coulombs = 1 decacoulomb (daC).

Decahydrate. A crystalline solid which is having ten molecules of water of crystallization per molecule of compound.

Decay Constant. Radioactive decay occurs according to the exponential rate law $N = N_0 e^{-\lambda t}$ where N_0 is the number of atoms present at time. zero and N is the number of atoms present at time t, λ is the decay constant, which is equal to $0.693/t_{\frac{1}{2}}$, $t_{\frac{1}{2}}$ being the half-life of the radioactive species, and has a characteristic value for a particular nuclide.

Deci. Symbol d. A prefix used in the metric system to denote one-tenth. For example, 0.1 coulomb = 1 decicoulomb (dC).

Decomposition. Chemical reaction in which a compound breaks down into simpler compounds or into elements.

Decomposition Voltage. The smallest voltage which will cause the electrolysis of an electrolyte. The decomposition voltage or potential depends both upon the nature of the electrolyte and of the electrodes.

Defect. Refers to an irregularity in the ordered arrangement of particles in a crystal lattice. There are two main types of defect in crystals. *Point defects* occur at single lattice points and there are two types. A *vacancy* is a missing atom ; *i.e.* a vacant lattice point. Vacancies are sometimes called *Schottky defects*. An *interstitial* is an atom that is in a position that is not a normal lattice point. If an atom moves off its lattice point to an interstitial position the result (vacancy plus interstitial) is a *Frenkel defect*. All solids above absolute zero have a number of point defects, the concerntration of defects depending on the temperature. Point defects can also be produced by strain or by irradiation.

Dislocations (or *line defects*) are also produced by strain in solids. They are irregularities extending over a number of lattice points along a line of atoms.

Defect Structure. Structures in which there are irregularities in the lattice. Defects may be of many types but include the non-alignment of crystalities, orientational disorder (*e.g.* of molecules or ions), vacant sites with the migrated atom at the surface (Schottky defect), vacant sites with an interstitial atom (Frenkel defects), non-stoicheiometry. Crystals exhibiting defects show anomalous physical properties, particularly density, colour, electrical conductivity. Applications of defect solids include catalysis, semi-conductors, luminescent materials.

Degenetacy. Having the same energy ; particularly applied to wave-functions. In a transition element gaseous atom the five d-orbitals are degenerate. In a crystal field the degeneracy is removed.

Degenerate Orbitals. See degeneracy.

Degree. A division on a temperature scale.

Degrees of Freedom 1. The number of independent parameters required to specify the configuration of a system. This concept is applied in the kinetic theory of specify the number of independent ways in which an atom or molecule can take up energy. There are however various sets of parameters that may be chosen, and the details of the consequent theory vary with the choice.

In a monatomic gas, such as helium or argon, the atoms have three translational degrees of freedom (corresponding to motion in three mutually perpendicular directions). The mean energy per atom for each degree of freedom is $kT/2$, where k is the Boltzmann constant ; the mean energy per atom is thus $3kT/2$.

A diatomic gas also has two rotational degrees of freedom (about two axes perpendicular to the bond) and one vibrational degree. The rotations also each contribute $kT/2$ to the average energy. The vibration contributes kT ($kT/2$ for kinetic energy and $kT/2$ for potential energy). Thus, the average energy per molecule for a diatomic molecule is $3kT/2$ (translation) $+ kT$ (rotation) $+ kT$ (vibration) $= 7kT/2$.

Linear triatomic molecules also have two significant rotational degrees of freedom ; non-linear molecules have three. For non-linear polyatomic molecules, the number of vibrational degrees of freedom is $3N-6$, where N is the number of atoms in the molecule.

1. The molar energy of a gas is the average energy per molecule multiplied by the Avogadro constant. For a monatomic gas it is $3\,RT/2$, etc.

2. The independent physical quantities (*e.g.*, pressure, temperature, etc.) that define the state of a given system.

3. In statistical mechanics (*e.g.*, the theory of specific heats of gases) a degree of freedom means an independent mode of absorbing energy by movement of atoms. Thus a monatomic gas has three translational degrees of freedom. Polyatomic molecules have in addition vibrational and rotational degrees of freedom.

Degree of Hydrolysis. The degree of hydrolysis of a salt is defined as the fraction of the total salt hydrolysed by water. Thus, if in a solution of the salt AB, 90% of it is hydrolysed to yield the base AOH and the acid BH, the degree of hydrolysis is 0.90. It may be expressed also as a percentage, *i.e.* 90%.

Deliquenscence. The absorption of water from the atmosphere by a hygroscopic solid to such an extent that a concentrated solution of the solid eventually forms.

Deliquescent. Describing a solid compound that absorbs water from the atmosphere, eventually forming a solution. See also hygroscopic.

Deliquescent Substance. A material showing deliquescence.

Demulsification. Emulsions of oil aod water formed in various operations may require special techniques to separate the two liquid phases. This process of demulsification can be effected by either physical or chemical means. The chemical methods depend on the opposing action of different emulsifiers on one another. Some emulsions can be broken by mechanical treatment while others can be broken electrically by a high voltage current. Some emulsions coagulate in the presence of electrolytes.

Dendrite. A crystal that has branched in growth into two parts. Crystals that grow in this way (*dendritic growth*) have a branching treelike appearance.

Dendritic Growth. Growth of crystals in a branching habit.

Densitometer. An appartus for measuring the intensities of *e.g.* lines on a photographic plate. Used in X-ray analysis and spectrometric analysis.

Density. May be defined as the mass of a substance per unit of volume. In SI units it is measured in kg m^{-3}. *See also* relative density ; vapour density.

Depolarizer. A substance which is used in a voltaic cell to prevent polarization. Hydrogen bubbles forming on the electrode can be removed by an oxidizing agent, such as manganese (IV) oxide.

Depolarization. The prevention of polarization in a primary cell. For example, maganese (IV) oxide (the *depolarizer*) is placed around the positive electrode of a Leclanche cell to oxidize the hydrogen released at this electrode.

Depression of Freezing Point. The reduction in the freezing point of a pure liquid when another substance is dissolved in it. It is a colligative property—*i.e.*, the lowering of the freezing point in proportional to the number of dissolved particles (molecules or ions), and does not depend on their nature.

The depression caused by 1 mole of solute in 1 litre of solvent is called the molal depression constant (K_f). (K_f) has a define value for a particular solvent.

Solvent	*Molecular depression constant (°C/mole)*
water	1.86
ethanoic acid	3.90
benzene	4.90
cyclohexane	20.5

The depression depends only on the concentration and is independent of solute composition. The proportionality constant, K_f, is called the *freezing point* constant or sometimes the *cryoscopic constant*. $\Delta t = K_f C_m$, where Δt is the lowering of the temperature and C_m is the molal concentration ; the units of K_f are kelvins kilograms moles^{-1} (K kg mol^{-1}). Although closely related to the property of boiling-point elevation, the cryogenic method can be applied to measurement of relative molecular mass with considerable precision. A known weight of pure solvent is slowly frozen, with stirring, in a suitable cold bath and the freezing temperature measured using a Backmann thermometer. A known weight of solute of known molecular mass is introduced, the solvent thawed out, and the cooling process and measurement repeated. The addition is repeated several times and in average value of K_f for the solvent obtained by plotting Δt against C_m. The whole process is then repeated using the unknown solute and its relative molecular mass determined using the value of K_f previously obtained.

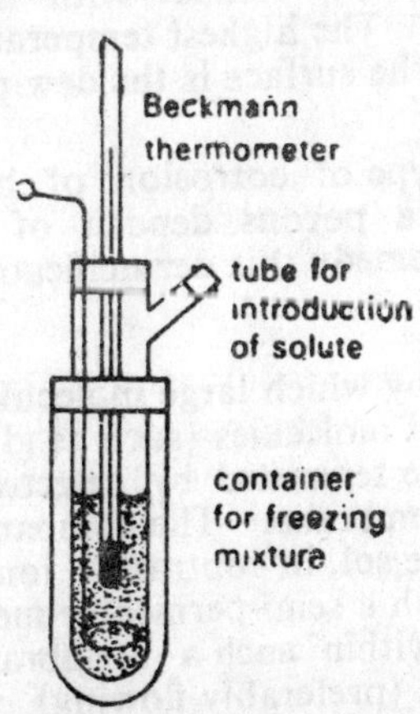

Apparatus for measuring depression of freezing point.

Derived Unit. A unit which is defined in terms of base units, and not directly from a standard value of the quantity it measures. For example, the newton is a unit of force defined as a kilogram metre second^{-2} (kg ms^{-2}).

Desorption. The removal of absorbed atoms, molecules, or ions from a surface.

Dewar Flask. A vessel which is used for storing hot or cold liquids so that they maintain their temperature independently of the surroundings. Heat transfer to the surroundings gets reduced to a minimum : the walls of the vessel consists of two thin layers of glass (or, in large vessels, steel) separated by a vacuum to reduce conduction and convection ; the inner surface of a glass vessel has been silvered to reduce radiation ; and the vessel is stoppered to prevent evaporation. It was devised around 1872 by the British physicist Sir James Dewar (1842-1923) and is also known by its first trade name *Thermos flask.*

Dew Point. Refers to the temperature at which liquid first appears when a mixture of vapours is cooled, assuming the system is at equilibrium.

Dew Point Hygrometer. An instrument which is used for the determination of dew point. It consists of a highly polished surface which is cooled gradually, usually by the controlled evaporation of ether, in contact with the vapour whose dew point is required. The highest temperature at which condensation occurs on the surface is the dew point.

Dezincification. A type of corrosion of brasses in which Zn is removed to leave a porous deposit of copper. Sn or As are incorporated to remedy this dezincification in admiralty metal and naval brass.

Dialysis. A method by which large molecules (such as starch or protein) and small molecules (such as glucose or amino acids) in solution may be separated by selective diffusion through a semipermeable membrane. The comparatively large particles (1-100pm) of the sol, in contrast to ions and molecules, cannot diffuse through a semi-permeable membrane. By enclosing the impure sol within such a membrane (dialyser), which is supported in pure (preferably flowing) water, impurities of a crystalloid nature are gradually removed. The function of the kidneys in mammals is to dialyse the blood, removing impurities such as urea from the haemoglobin sol.

Diamond. An allotrope of carbon. It is the herdest naturally occurring substance and is used for jewellery and, industrially, for drilling equipment. Each carbon atom is surrounded by four equally spaced carbon atoms arranged tetrahedrally. The carbon atoms form a three-dimensional network with each carbon-carbon bond equal to 0.154 nm and at an angle of 109.5° with its neighbours. In diamonds millions of atoms are covalently bonded to form a giant molecular structure, the great strength of which results from the strong covalent bonds.

Diaphragm Cell. A cell for electrolysis of brine in which the anode and cathode compartments are separated by a diaphragm.

Dielectric Constant. The force F between two electric charges e, separated by a distance r in a vacuum, is given by $F=e^2/r^2$. In any other medium this relationship becomes $F=e^2/Dr^2$ where D is the dielectric constant of the medium. The dielectric constant is a measure of the polarity of the medium, some typical values being 1.00 (air) ; 1.013 (water as steam) ; 1.874 (hexane) ; 5.94 (chlorobenzene) ; 15.5 (liquid ammonia) and 80 (water at 20°C).

Dieterici's Equation. A modification of van der Waal's equation, in which account is taken of the pressure gradient at the boundary of the gas. It is written

$$p(v-b)=RTe^{-a/RTV}$$

where a and b are constants characteristic of each gas ; at relatively low pressures it is identical with van der Waals' equation.

Diffraction Pattern. If a beam of X-rays or electrons is passed through a crystalline solid and allowed to impinge on a photographic plate, a pattern consisting of a more or less symmetrical arrangement of spots of rings is produced. With non-crystalline materials the rings are diffuse. The pattern is due to the diffraction of the X-rays or electrons in a manner analogous to the diffraction of light by a diffraction grating. Reflected X-rays or electrons from crystalline materials also give diffraction patterns. The diffraction patterns are used in structure determination.

Diffusion. 1. Meovement of a gas, liquid, or solid as a result of the random thermal motion of its particles (atoms or molecules). A drop of ink in water, for example, will slowly spead throughout the liquid. Diffusion in solids occurs very slowly at normal temperatures. In gases, all the components are perfectly miscible with each other and mixing ultimately becomes nearly uniform, though slightly affected by gravity (see also Graham's law). The diffusion of a solute through a solvent to produce a solution of uniform concentration is slower, but otherwise very similar to the process of gaseous diffusion. In solids, however, diffusion occurs very slowly at normal temperatures. 2. The passage of elementary particles through matter when there is a high probability of scattering and a low probability of capture.

Dihedral (Dihedron). An angle which is formed by the intersection of two planes (*e.g.*, two faces of a polyhedron.) The *dihedral angle* is the angle formed by taking a point on the line of intersection and drawing two lines from this point, one in each plane, perpendicular to the line of intersection.

Dilatometer. An apparatus for measuring small changes of volume of a liquid, solution or solid immersed in a liquid. It usually consists of a cylindrical glass bulb with attached capillary tube, and change of volume of the contents of the bulb is noted by observing the movement of the meniscus in the capillary. Dilatometry is useful for determining transition temperatures, rates of reaction, polymerization, etc.

Dilute. Denoting a solution in which the amount of solute is low relative to that of the solvent. The term is always relative and includes dilution at trace level as well as the common term 'bench dilute acid', which usually means a 2M solution. *Compare* concentrated.

Dilution. The volume of solvent in which a given amount of solute is dissolved.

Dilution Law. See Ostwald's dilution law.

Dimorphism. See polymorphism.

Discharge Tube. A vessel, usually glass, which contains two metal electrodes and in which the passage of electricity occurs through a gas or vapour (usually at low pressure).

Dislocation. See defect.

Disperse Phase. See colloid.

Dispersion Force. A weak type of intermolecular force. See van der Waals forces.

Displacement Pump. A commonly used device which is used for transporting liquids and gases around chemical plants. It works on the principle of the bicycle pump : a piston raises the pressure of the fluid and, when it is high enough, a valve opens and the fluid is discharged through an outlet pipe. As the piston moves back the pressure falls and the cycle continues. Displacement pumps can be used to generate very high pressures (*e.g.*, in the synthesis of ammonia) but because of the system of valves, they have been more expensive than other types of pump.

Displacement Reaction. A chemical reaction in which an atom or group displaces another atom or group from a molecule. A common example is the displacement of hydrogen from acids by metals :

$$Zn + 2HCl \rightarrow ZnCl_2 + H_2$$

Disproportionation. Refers to a type of chemical reaction in which the same compound gets simultaneously reduced and oxidized, For example, copper(I) chloride disproportionates thus :

$$2CuCl \rightarrow Cu + CuCl_2$$

The reaction involves oxidation of one molecule

$$Cu^{I} \rightarrow Cu^{II} + e$$

and reduction of the other

$$Cu^{I} + e \rightarrow Cu$$

The reaction of halogens with hydroxide ions is another example of a disproportionation reaction, for example

$$Cl_2(g) + 2OH^-(aq) \rightleftharpoons Cl^-(aq) + ClO^-(aq) + H_2O(l)$$

Dissociation. The process whereby a molecule is split into simpler fragments which may be smaller molecules, atoms, free radicals of ions. Particularly used in connection with photochemical, thermal and ionic (electrolytic) dissociation. The extent of dissociation is measured by the dissociation constant, K. For the process $AB \rightleftharpoons A+B$

$$K=\frac{[A][B]}{[AB]}$$

where the square brackets denote the concetrations (activities) of the species.

Dissociation Constant. The equilibrium constant of a dissociation reaction. For example, the dissociation constant of a reaction :

$$AB=A+B$$

is given by :

$$K=[A][B]/[AB]$$

Often the degree of dissociation is used—the fraction (α) of the original compound that has dissociated at equilibrium. For an original amount of AB of *n* moles in a volume V, the dissociation constant is given by :

$$K=\alpha^2 n/(1-\alpha)V$$

Note that this expression is for dissociation into two molecules.

Acid dissociation constants (or *acidity constants*, symbol : K_a) are dissociation constants for the dissociation into ions in solution.

$$HA+H_2O \rightarrow H_3O^+ + A$$

The concentration of water can be taken as unity, and the acidity constant is given by :

$$K_a=[H_3O^+][A^-]/[HA]$$

The acidity constant is a measure of the strength of the acid. Base dissociation constants (K_b) are similarly defined.

The expression :

$$K=\alpha^2 n/(1-\alpha)V$$

applied to an acid is known as *Ostwald's dilution law.* In particular if α is small (a weak acid) then $K=\alpha^2n/V$, or $\alpha=C\sqrt{V}$, where C is a constant. The degree of dissociation is then proportional to the square root of the dilution.

Dissociation Pressure. If a solid compound dissociates to give one or more gaseous products, the dissociation pressure is the pressure of gas in equilibrium with the solid at a given temperature. For example, when calcium carbonate is maintained at a constant high temperature in a closed container, the dissociation pressure at that temperature is the pressure of carbon dioxide from the equilibrium

$$CaCO_3(s) \rightleftharpoons CaO(s) + CO_2(g)$$

Distillation. Refers to the process of boiling a liquid and condensing the vapour. Distillation is used to purify liquids or to separate components of a liquid mixture. See also destructive distillation, fractional distillation, steam distillation, vacuum distillation.

Distilled Water. Water purified by distillation so as to free it from dissolved salts and other compounds. Distilled water in equilibrium with the carbon dioxide in the air has a conductivity of about $0.8+10^{-6}$ siemens cm^{-1}. Repeated distillation in a vacuum can bring the conductivity down to 0.043×10^{-6} siemens cm^{-1} at 18°C (sometimes called *conductivity water*). The limiting conductivity is due to self ionization :

$$H_2O \rightleftharpoons H^+ + OH^-.$$

Diverse Salt Effect. Effect of a salt without common ions on precipitation because of the effect of the ions on the activity of the precipitating ions.

Double Refraction. The property, possessed by certain crystals (notably calcite), of forming two refracted rays from a single incident ray. The *ordinary ray* obeys the normal laws of refraction. The other refracted ray, called the *extraordinary ray*, follows different laws. The light in the ordinary ray is polarized at right angles to the light in the extraordinary ray. Along an optic axis the ordinary and extraordinary rays travel with the same speed. Some crystals, such as calcite, quartz, and tourmaline, have only one optic axis ; they are *uniaxial crystals*. Others, such as mica and selenite, have two optic axes ; they are *biaxial crystals*. The phenomenon is also known as ***birefringence*** and the double-refracting crystal as a ***birefringent crystal***. See also polarization.

Doublet. Term used in *n.m.r.* spectroscopy to denote a signal split into an equal doublet by coupling to a further nucleus with spin ½. Also used in group theory to denote particular energy states, *e.g.*, with s=2.

Dropping Mercury Electrode, DME. An electrode consisting of a column of mercury passing through a fine capillary and emerging in the solution as fine drops to give a continuously fresh surface. Used in polarography.

Dry Cell. A voltaic cell in which the electrolyte is in the form of a jelly or paste. The Leclanche dry cell is extensively used for flashlights and other portable applications. See also Leclanche cell.

Dry Ice. Solid carbon dioxide used as a refrigerant. It is convenient because it sublimes at −78°C (195 K) atst andard pressure rather than melting.

Dulong and Petit's Law. For a solid element the product of the relative atomic mass and the specific heat capacity is a constant equal to about 25J mol^{-1} K^{-1}. Formulated in these terms in 1819 by the French scientists Pierre Dulong (1785-1838) and A.T. Petit (1791-1820), the law in modern terms states : the molar heat capacity of a solid element is approximately equal to 3R, where R is the gas constant.

The law applies only to elements with simple crystal structures at normal temperatures. At lower temperatures the molar heat capacity falls with decreasing temperature (it is proportional to T^3). Molar thermal capacity was formerly called *atomic heat*—the product of the atomic weight and specific thermal capacity.

Dynamic Equilibrium. See equilibrium.

Dyne. Symbol : dyn. The former unit of force used in the c.g.s. system. It is equal to 10^{-5} N.

E

Ebullioscopic Constant. See elevation of boiling point.

Ebullition. The boiling or bubbling of a liquid.

Edison Cell. See nickel-iron accumulator.

Efficiency. Symbol : η. A measure which is used for processes of energy transfer ; the ratio of the useful energy produced by a system or device to the energy input. For a reversible heat engine, the efficiency has been given by

$$\eta=(T_1-T_2)/T_1$$

T_1 being the source temperature and T_2 the 'sink' temperature.

Efflorescence. Refers to the process in which a crystalline hydrated solid loses water of crystallization to the air. A powdery deposit is gradually formed.

Effusion. The passage of a gas through an orifice which has a diameter smaller than the mean free path of the gas molecules. The rate of effusion is proportional to the area of the orifice and to the mean velocity of the molecules (and hence to the reciprocal of the square root of the density or molecular weight of the gas). This principle is employed experimentally for investigating low vapour pressures and the molecular weight of gaseous species (*e.g.* in high temperature equilibria) ; the apparatus used is called a Knudsen cell.

Einstein. A unit of radiant energy. If every molecule in a gram molecule of a substance absorbs one quantum of light, the total energy absorbed is Lhv, where L is the Avogadro number, that is, the number of molecules in one gram molecule, h is Planck's constant, and v the frequency of the light. $L=6.06\times10^{23}$; $h=6.548\times10^{-27}$ erg sec., and therefore for red light of frequency 4×10^{14}, one einstein$=1.588\times10^{11}$ ergs.

Einstein Equation. 1. The mass-energy relationship announced by Einstein in 1905 in the form $E=mc^2$, where E is a quantity of energy, m its mass, and c is the speed of light. It presents the concept that energy possesses mass. 2. The relationship $E_{max}=hf-W$. where E_{max} is the maximum kinetic energy of electrons emitted in the photo-emissive effect, h is the Planck constant, f the frequency of the incident radiation, and W the work function of the emitter. This is also written $E_{max}=hf-\phi e$, where e is the electronic charge and ϕ a potential difference, also called the work function. (Sometimes W and ϕ are distinguished as *work function energy* and *work function potential.*) The equation can also be applied to photoemission from gases, when it has the form : $E=hf-I$, where I is the ionization potential of the gas.

Einstein's Law of Photochemical Equivalence. Before a chemical reaction can be induced by light, the light must be absorbed by some of the reactant molecules. Einstein proposed that each reacting molecule must be excited by absorption of one quantum of light. This principle is not easy to test, however, since secondary reactions occurring after the primary reaction may cause the destruction of many more of the molecules of the absorbing substance. In addition, excited molecules may lose energy by such processes as fluorescence or by undergoing deactivating collisions. In consequence the yield of reaction products may only be a fraction of that calculated from Einstein's law. The law is, however, considered to hold for the primary act of light absorption.

Electrical Double Layer. An electrical double layer is generally formed at the interface between two phases. One phase acquires a net positive charge and the other a net negative charge. The first quantitative investigation of this phenomenon was by Helmholtz (1879) who, using a parallel plate condenser as his model, assumed that there existed a compact double layer at the ionized surface. This was later modified by Gouy who suggested that although the inner ionic layer of the double layer is rigid (owing to absorbed or structural ions) the outer layer is more diffuse and extends a few mμ from the surface. Thus a negatively-charged dropping mercury electrode would be surrounded by an ionic sheath in which the proportion of positive ions decreases and that of negative ions increases with increasing distance from the interface. Stern (1924) improved on the model by combining the essential characters of both the rigid and diffuse double layer theories.

Electrochemical Cell. See cell.

Electrochemical Equivalent. Symbol *z*. The mass of a given element liberated from a solution of its ions in electrolysis by one coulomb of charge. See Faraday's laws (of electrolysis).

Electrochemistry. The branch of chemistry concerned with electrolysis and other similar phenomena occurring when a current is passed through a solution of an electrolyte, or concerned with the behaviour of ions in solution and the properties shown by these solutions.

Electrode. Any part of an electrical device or system that emits or collects electrons or other charge carriers. An electrode may also be used to deflect charged particles by the action of the electrostatic field that it produces. See *also half cell.*

Electrodeposition. Refers to the process of depositing a layer of solid (metal) on an electrode by electrolysis. Positive ions in solution gain electrons at the cathode and are deposited as atoms. Copper, for instance, can be deposited on a metal cathode from copper sulphate solution.

Electrode Potential. The potential difference produced between the electrode and the solution in a half cell. It is not possible to measure this directly since any measurement involves completing the circuit with the electrolyte, thereby introducing another half cell. *Standard electrode potentials* E are defined by measuring the potential relative to a standard hydrogen half cell using 1.0 molar solution at 25°C. The convention is to designate the cell so that the oxidized form is written first.

Electrode potentials are also called *reduction potentials*. The sign given to the electrode potential is arbitrary ; values are usually quoted as 'oxidation' potentials, *i.e.*, the electrode potential for the process $M \rightarrow M^{n+}$. A positive electrode potential indicates that the oxidation process tends to occur spontaneously. A negative oxidation potential for $M \rightarrow M^{n+}$ indicates that the process $M^{n+} \rightarrow M$ tends to occur spontaneously.

If an appreciable current flows between the electrode and the solution, thus disturbing the reversible thermodynamic equilibrium conditions, the electrode is said to be polarized and the system is then operating under irreversible conditions.

It is not, however, possible to measure electrode potential for an isolated half cell—any measurement requires a circuit, which sets up another half cell in the solution. Therefore, electrode potentials (or *reduction potentials*) are defined by comparison with a hydrogen half cell, which is connected to the half cell under investigation by a salt bridge. The e.m.f. of the full cell can then be measured.

In referring to a given half cell the more reduced form is written on the right for a half-cell reaction. For the half cell Cu^{2+} | Cu, the half-cell reaction is a reduction :

$$Cu^{2+}(aq) + 2e \rightarrow Cu$$

The cell formed in comparison with a hydrogen electrode is :

$$Pt(s)H_2(g) \mid H^+(aq) \mid Cu^{2+}(aq) \mid Cu$$

The e.m.f. of this cell is +0.34 volt measured under standard conditions. Thus, the standard electrode potential (symbol : E^{θ}) is +0.34 V for the half cell Cu^{2+} | Cu. The standard conditions are 1.0 molar solutions of all ionic species, standard pressure, and a temperature of 298 K.

Half cells can also be formed by a solution of two different ions (*e.g.* Fe^{2+} and Fe^{3+}). In such cases, a platinum electrode is used under standard conditions.

Electrodialysis. A method of obtaining pure water from water containing a salt, as in desalination. The water to be purified is fed into a cell containing two electrodes. Between the electrodes is placed an array of semipermeable membranes alternately semipermeable to positive ions and negative ions. The ions tend to segregate between alternate pairs of membranes leaving pure water in the other gaps between membranes. In this way, the feed water is separated into two streams : one of pure water and the other of more concentrated solution.

Electrodispersion. When an arc is struck between two metal electrodes under the surface of a liquid, particles of the metal are torn off and these remain colloidally dispersed in the liquid. Using this method colloidal dispersions of most metals were prepared by Bredig. A more refined electrodispersion process is obtained if the arc is maintained within a silica tube and the vaporized metal blown through a small hole in front of the arc into the dispersion medium using a stream of inert gas.

Electrokinetic Potential. Frequently called the zeta (ξ) potential, it is the potential difference across the diffuse part of a double layer, *i.e.*, between the rigid solution layer and the mobile part of the solution adjacent to the bulk solution. See electrical double layer.

Electrokinetics. Electrophoresis is one of several effects, collectively known as electrokinetic phenomena, which are met in systems with electrical double layers at the interface between a solid and a liquid (generally aqueous) or between one liquid and another. The other effects are electro-osmosis, streaming potential and sedimentation potential. They all arise from a partial separation of the fixed and mobile parts of the electrical double layer.

Electroluminescence. See luminescence.

Electrolysis. The process of decomposing a substance, usually in solution or as a melt, by the passage of an electric current.

The current is conducted by migration of ions—positive ones (cations) to the cathode (negative electrode), and negative ones (anions) to the anode (positive electrode). Reactions take place at the electrodes by transfer of electrons to or from them.

In the electrolysis of water (containing a small amount of acid to make it conduct adequately) hydrogen gas is given off at the cathode and oxygen is evolved at the anode. At the cathode the reaction is :

$$H^+ + e^- \rightarrow H$$
$$2H \rightarrow H_2$$

At the anode :

$$OH^- \rightarrow e^- + OH$$
$$2OH \rightarrow H_2O + O$$
$$2O \rightarrow O_2$$

In certain cases the electrode material may dissolve. For instance, in the electrolysis of copper(II) sulphate solution with copper electrodes, copper atoms of the anode dissolve as copper ions

$$Cu \rightarrow 2e^- + Cu^{2+}$$

Electrolyte. A liquid that conducts electricity as a result of the presence of positive or negative ions. Electrolytes are molten ionic compounds or solutions containing ions, *i.e.*, solutions of ionic salts or of compounds that ionize in solution. Liquid metals, in which the conduction is by free electrons, are not usually regarded as electrolytes. Electrolytes with dissociation constants greater than about 10^{-2} are called strong electrolytes.

Electrolyte Dissociation. The process of the formation of ions in solution from an added solute. In the case of solid solutes with an ionic lattice (*e.g.* NaCl) there is probably complete dissociation into ions in dilute solution. For other types of solute (*e.g.* ethanoic acid) there is an equilibrium between the ions and the undissociated molecules.

Electrolytic. Relating to the behaviour or reactions of ions in solution.

Electrolytic Cell. See cell, electrolysis.

Electrolytic Oxidation. An oxidation process effected by electrolysis.

Electrolytic Reduction. A reduction process effected by electrolysis.

Electromagnetic Radiation. Around the end of the nineteenth century it was realized that certain optical experiments could best be understood by considering that light was produced by the oscillating motion of an electric charge. This oscillation causes the associated electric field to change periodically and also produces an oscillating magnetic field. These oscillating electric and magnetic disturbances are propagated through space as waves of electromagnetic radiation.

The different forms of electromagnetic radiation, which have different wavelengths and energies, include visible, infra-red and ultra-violet light as well as radiowaves and X-rays. All forms of electromagnetic radiation are propagated through a vacuum with the same velocity of 2.98×10^8 m sec^{-1}. Electromagnetic radiation interacts with molecules according to the frequency (energy of the radiation) u.v.—visible—near i.r. (electronic transitions), i.r. (vibrations), microwave (rotations), radio frequencies (nuclear spins), X-rays (removal of electrons).

Electrometer. An instrument for detecting and measuring the magnitude of an electric charge.

Electrometric Titration. Analytical method for following titrations by observing the e.m.f. of an inert electrode immersed in the solution. Of special use in titrating coloured solutions.

Electromotive Force (**e.m.f.**). Refers to the greatest potential difference that can be generated by a particular source of electric current. In practice this may be observable only when the source is not supplying current, because of its internal resistance.

Electron. An elementary particle with a rest mass of 9.109558×10^{-31} kg and a negative charge of 1.602192×10^{-19} coulomb. Electrons are present in all atoms in groupings called shells around the nucleus, when they are detached from the atom they are called *free electrons*. The antiparticle of the electron is the *positron*.

Electron Affinity. Symbol A. The energy change occurring when an atom or molecule gains an electron to form a negative ion. For an atom or molecule X, it is the energy released for the electron-attachment reaction

$$X(g) + e \rightarrow X^-(g)$$

Often this is measured in electronvolts. Alternatively, the molar enthalpy change, ΔH, can be used.

Electron Capture. 1. The formation of a negative ion by an atom or molecule when it acquires an extra free electron. 2. A radioactive transformation in which a nucleus acquires an electron from an inner orbit of the atom, thereby transforming, initially, into a nucleus with the same mass number but an atomic number one less than that of the original nucleus (capture of the electron transforms a proton into a neutron). This type of capture is accompanied by emission of an X-ray photon or Auger electron as the vacancy in the inner orbit is filled by an outer electron.

Electron-Deficient Compound. A compound in which there are fewer electrons forming the chemical bonds than required in normal electron-pair bonds. See borane.

Electron Density. In the investigation of crystal structure by X-ray diffraction the electron density distribution which is calculated is mapped out and contour diagrams made by drawing lines of equal electron density are constructed. Using such diagrams accurate bond lengths may be derived.

In theories of bonding the term is often used to indicate the probability of finding an electron at a particular point.

Electromotive Series (Electrochemical Series). A series of chemical elements arranged in order of their electrode potentials. The hydrogen electrode ($H^+ + e \rightarrow \frac{1}{2}H_2$) is taken as having zero electrode potential. Elements that have a greater tendency than hydrogen to lose electrons to their solution are taken as *electropositive* ; those that gain electrons from their solution are below hydrogen in the series and are called *electronegative*. The series shows the order in which metals replace one another from their salts ; electropositive metals will replace hydrogen from acids. The chief metals and hydrogen, placed in order in the series, are : potassium, calcium, sodium, magnesium, aluminium, zinc, cadmium, iron, nickel, tin, lead, hydrogen, copper, mercury, silver, platinum, gold.

Electret. A permanently electrified substance or body that has opposite charges at its extremities. Electrets resemble permanent magnets in many ways. An electret can be made by cooling certain waxes in a strong electric field.

Electro-Osmosis. If a direct current is passed through a tube of liquid which is fitted with two electrodes, one on either side of a diaphragm, the potential difference between the diaphragm, and the liquid manifests itself by a movement of the liquid towards one or other of the electrodes. This phenomenon, which is akin to electrophoresis, is termed electro-osmosis. It is modified by the presence of acids, bases and salts. The hydrogen-ion concentration at which there is no movement of liquid with respect to the membrane is termed the isoelectric point.

Electrophoresis. The migration of charged particles, colloidal particles or ions through a solution under an electric field. The movement can be observed directly or by observing a boundary, *e.g.*, a coloured boundary. Variation of pH can stop movement at the isoelectric point (when the charge on the species changes). In electro-osmosis there is a constant flow of liquid relative to a stationary surface.

There are various experimental methods. In one the sample is placed in a U-tube and a buffer solution added to each arm, so that there are sharp boundaries between buffer and sample. An electrode is placed in each arm, a voltage applied, and the motion of the boundaries under the influence of the field is observed. The rate of migration of the particles depends on the field, the charge on the particles, and on other factors, such as the size and shape of the particles. More simply, electrophoresis can be carried out using an adsorbent, such as a strip of filter paper, soaked in a buffer with two electrodes making contact. The sample is placed between the electrodes and a voltage applied. Different components of the mixture migrate at different rates, so the sample separates into zones. The components can be identified by the rate at which they move. This technique has also been called *electrochromatography.*

Electrophoresis is used in analysis, particularly in biochemical applications (ionography, zone electrophoresis, electrochromatography) for both identification and separation.

Electrophoretogram. The separated species on, *e.g.*, a column after electrophoresis.

Electroplating. The deposition of metals from solution in the form of a layer on other metals or, *e.g.*, plastics by passage of an electric current. A metallic article to be plated is made one electrode in a bath containing the other metal as aquo-ions or other complexes. Current density, pH, concentration, etc. all have a very marked effect on the adhesion and texture of the deposited metal. Among metals used for electroplating are Ag, Cr, Ni, Zn. Colloidal rubber, where the gel particles are charged, can be plated and the rubber coating subsequently vulcanized.

Electrostatic Precipitators. Plants for the removal of fine suspended matter from a gas that depend for their action on the ionization of the gas between two highly charged electrodes. The ions so formed attach themselves to the dispersed particles, conferring a charge, with the result that the latter then migrate to the appropriate electrode.

Electrostatic precipitators are relatively expensive and running costs high, but collection efficiencies are high also, even for very small particles. Material as small as 0.01 μ may be removed, and they are thus suitable for smokes and fumes.

Electrovalent Bond, Polar Bond. Bonding by electrostatic attraction.

Elements of Symmetry. The symmetry elements, centres, axes, planes of symmetry, present in a molecule, crystal lattice or crystal. Together with the arrangement of atoms the elements of symmetry spell out the space-group of a crystal or the point-group of a molecule.

Elevation of Boiling Point. An increase in the boiling point of a liquid when a solid is dissolved in it. The elevation is proportional to the number of particles dissolved (molecules or ions) and is given by $\Delta t = k_B C_M$ where C_M is the molal concentration of solute. The constant k_B is the *ebullioscopic constant* of the solvent and if this is known, the molecular weight of the solute can be calculated from the measured value of Δt. The elevation is measured by a Beckmann thermometer.

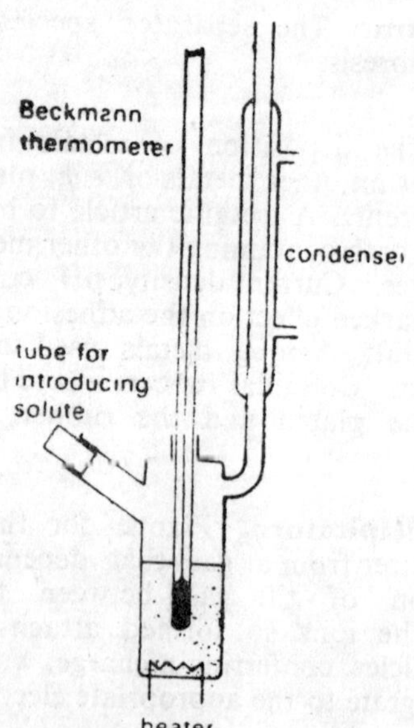

Apparatus for measuring elevation of boiling point

An accurately weighed amount of pure solvent is boiled until the temperature is steady, a known weight of solute of known molecular mass is quickly introduced, the boiling continued, and the elevation measured using a Beckmann thermometer. The process is repeated several times and the average value of K_b obtained by plotting Δt against C_M. The whole process is then repeated with the unknown material and its relative molecular mass obtained using the value of K_b previously obtained.

There are several disadvantages with this method and it is therefore used largely for demonstration purposes. The main problem is that the exact amount of solvent remaining in the liquid phase is unknown and varies with the rate of boiling. The theoretical explanation of the effect is identical to that for the lowering of vapour pressure ; the boiling points are those temperatures at which the vapour pressure equals the atmospheric pressure.

Elutriation. The process of suspending finely divided particles in an upward flowing stream of air or water to wash and separate them into sized fractions.

E.M.F. See electromotive force.

Emission Spectroscopy. An analytical technique in which atoms in excited states in flames or electrie arcs are examined by their emission spectra. Emission spectroscopy is also used to investigate the energy levels of atoms in ground and excited states.

Emission Spectrum. See spectrum.

Emulsification. The preparation of a suspension, or emulsion, of one liquid in another.

Emulsifier. (1) Another name for an emulsifying agent.

(2) Any machine for producing an emulsion. In some cases a satisfactory emulsion may be produced by the use of a simple mixer, such as a propeller or turbine agitator, but frequently it is necessary to employ a special machine. See colloid mills.

Emulsifying Agent. Dilute suspensions of oil in water behave as typical hydrophobic sols, and it is not possible to increase the concentration of oil unless a stabilizing material is added which decreases the interfacial tension. Such substances, known as emulsifying agents, are generally long-chain compounds containing a hydrophilic (carboxyl or sulphonate) group at one end of the molecule ; these become orientated at the interface, the hydrophilic end projecting into the water. There are also solid emulsifying agents (*e.g.*, carbon black) which possess widely different angles of contact for the two phases.

There are two types of emulsion possible for each system ; oil-in-water (O/W) and water-in-oil (W/O). Usually a given emulsifying agent will emulsify so as to form only one of these types. An emulsifying agent generally produces such an emulsion that the liquid in which it is most soluble forms the external phase. Thus the alkali metal soaps and hydrophilic colloids produce O/W emulsions, oil-soluble resins the W/O type.

The commonest emulsifying agents are the soaps, which have the disadvantage of coagulating in hard waters. For edible and pharmaceutical emulsions gums of various sorts are employed. Of recent years, large numbers of synthetic organic emulsifying agents have been developed. Many of these are sulphonates or quaternary ammonium compounds.

Emulsion. A colloid in which small particles of one liquid are dispersed in another liquid. Usually emulsions involve a dispersion of water in an oil or a dispersion of oil in water, and are stabilized by an *emulsifier*. Commonly emulsifiers are substances, such as detergents, that have lyophobic and lyophilic parts in their molecules.

The viscosity of an emulsion increases rapidly with the concentration of the disperse phase ; very often when the emulsifying agent is not efficient, increase in the concentration of the disperse phase above a certain value causes phase reversal, an O/W emulsion changing to W/O type. The presence of certain ionic substances also brings this about.

Emulsions have many industrial uses ; they are used widely as foods and pharmaceutical preparations, cosmetics, horticultural and insecticide sprays, as oil-bound water paints, as lubricants, for spraying of roads, etc.

Emulsion Stabilizers. Compounds which are weakly adsorbed on the grain surface of a photographic emulsion where they displace the chemical sensitizer and which prevent the appearance of fog. See antifoggants.

Endothermic. Describing a process in which heat is absorbed (*i.e.*, heat flows from outside the system, or the temperature falls). The dissolving of a salt in water, for instance, is often an endothermic process. *Compare* exothermic.

Energy. Refers to a measure of a system's ability to do work. Like work itself, it is measured in joules. Energy is conveniently classified into two forms : *potential energy* is the energy stored in a body or system as a consequence of its position, shape, or state (this includes gravitational energy, electrical energy, nuclear energy, and chemical energy) ; *kinetic energy* is energy of motion and is usually defined as the work that will be done by the body possessing the energy when it is brought to rest. For a body of mass m having a speed v, the kinetic energy is $mv^2/2$. The rotational kinetic energy of a body having an angular velocity ω is $I\omega^2/2$, where I is its moment of inertia.

The internal energy of a body refers to the sum of the potential energy and the kinetic energy of its component atoms and molecules.

Energy Levels. Energy taken up by a molecule, say by the adsorption of light, may cause an electron to move from an orbital of lower to one of higher energy ; it may increase the energy of rotation of the molecule as a whole, or may increase the energy of vibration of the nuclei of the constituent atoms relative to one another. The changes all occur in accordance with the quantum hypothesis, *i.e.*, the changes in energy are not continuous, but take place only in definite steps. Thus, every electronic orbital is associated with a specific energy value, or energy level. So the rotational and vibrational states have only certain restricted energy values, or energy levels. Diagrammatically, these are usually represented as a series of horizontal, parallel straight lines separated by distances proportional to the differences in energy of the various states.

Energy Profile. Refers to a diagram that traces the changes in the energy of a system during the course of a reaction.

Energy profiles could be obtained by plotting the potential energy of the reacting particles against the reaction coordinate. To obtain the reaction coordinate the energy of the total interacting system is plotted against position for the molecules. The *reaction coordinate* is the pathway for which the energy is a minimum.

The energy profile for the hydrogenation of an alkene with and without a catalyst is having certain interesting features : the activation energy without the catalyst is very much larger than that with the catalyst and any point along the horizontal axis represents the changes that the colliding particles experience as atoms are separated in bond breaking and as atoms are brought together in bond formation.

Engler Viscometer. An instrument used in some European countries for determination of viscosity of oils and other fluids. Times are measured for flow of oil and distilled water through a standard orifice at selected temperatures. Results are expressed in degrees Engler (°E), the ratio of efflux time for oil to effluximet for water.

Enantiotropy. The reversible transformation of one form or one allotrope into another at a definite transition temperature, *e.g.*, red $HgI_2 \rightleftharpoons$ yellow HgI_2 at 126°C.

Enthalpy. Symbol H. A thermodynamic property of a system defined by $H=E+pV$, where H is the enthalpy, E is the internal energy of the system, p its pressure, and V its volume. In a chemical reaction carried out in the atmosphere the pressure remains constant and the enthalpy of reaction, ΔH, is equal to $\Delta E+p\Delta V$. For an exothermic reaction ΔH is taken to be negative.

For any process which occurs at constant pressure the heat absorbed or evolved is equal to the enthalpy change if the only work is pressure/volume work, *i.e.*, $\Delta H=\Delta E+P\Delta V$ for a process at constant pressure. In chemical reactions the enthalpy change (ΔH) is related to the changes in free energy (ΔG) and entropy (ΔS) by the equation $\Delta G=\Delta H-T\Delta S$. Enthalpy is usually expressed in units of kilojoules per mole.

Entrainment. The carrying forward by a stream of gas or vapour of fine liquid droplets. In distillation columns the transport of liquid from one plate to the next by this mechanism can have an adverse effect on separation efficiency.

Entropy. Symbol S. A measure of the unavailability of a system's energy to do work ; an increase in entropy is accompanied by a decrease in energy availability. When a system undergoes a reversible change the entropy (S) changes by an amount equal to the energy (Q) absorbed by the system divided by the thermodynamic temperature (T) at which the energy is absorbed, *i.e.*, $\Delta S = \Delta Q/T$. However, all real processes are to a certain extent irreversible changes and in any closed system an irreversible change is always accompanied by an increase in entropy.

In a wider sense entropy can be interpreted as a measure of a system's disorder ; the higher the entropy the greater the disorder. As any real change to a closed system tends towards higher entropy, and therefore higher disorder, it follows that the entropy of the universe (if it can be considered a closed system) is increasing and its available energy is decreasing. This increase in the entropy of the uninverse is one way of stating the second law of thermodynamics.

In any real (irreversible) change in a closed system the entropy increases. Although the total energy of the system has not changed (first law of thermodynamics) the available energy is less—a consequence of the second law of thermodynamics.

The concept of entropy has been widened to take in the general idea of disorder—the higher the entropy, the more disordered the system. For instance, a chemical reaction involving polymerization may well have a decrease in entropy because there is a change to a more ordered system. The 'thermal' definition of entropy is a special case of this idea of disorder—here the entropy measures how the energy transferred is distributed amongst the particles of matter.

Enzyme. A protein, with a molecular weight between 10^5 and 10^7, that catalyses a specific biochemical reaction. An enzyme mediates the conversion of one substance (the *substrate*) to another (the *product*) by combining with the substrate to form an intermediate complex. The enzyme molecule is very much larger than the substrate molecule. Enzymes differ from inorganic catalysts in the following ways :

(1) they have a high degree of specificity—enzymes can even distinguish between enantiomorphs ;

(2) an increase in temperature can cause a decrease in the rate of reaction as a result of the enzyme being denatured ;

(3) they are destroyed by too great a change in pH ;

(4) they are inactivated by low concentrations of heavy-metal ions.

Enzymes act by lowering the overall activation energy of a reaction sequence by involving a series of intermediates, or a mechanism, different from the spontaneous uncatalysed reaction.

Most enzymes work best within a narrow pH range and an susceptible to a wide variety of compounds which inhibit or sometimes promote the activity. The majority of enzymes work most efficiently at around 40°C and at higher temperatures are rapidly destroyed.

Equation of State. Refers to an equation that relates the pressure p, volume V, and thermodynamic temperature T of an amount of substance n. The simplest is the ideal gas law :

$$pV = nRT,$$

where R is the universal gas canstant. Applying only to ideal gases, this equation takes no account of the volume occupied by the gas molecules (according to this law if the pressure is infinitely great the volume becomes zero), nor does it take into account any forces between molecules. A more accurate equation of state would therefore be

$$(p+k)(V-nb) = nRT,$$

where k is a factor that reflects the decreased pressure on the walls of the container as a result of the attractive forces between particles, and nb is the volume occupied by the particles themselves when the pressure is infinitely high. In the *van der Waals equation of state*, proposed by the Dutch physicist J.D. van der Waals (1837-1923),

$$k = n^2a/V^2,$$

where a is a constant. This equation more accurately reflects the behaviour of real gases ; several others have done better but are more complicated.

Equilibrium. In a reversible chemical reaction :

$$A+B=C+D$$

The reactants are forming the products :

$$A+B \rightarrow C+D$$

which also react to give the original reactants :

$$C+D \rightarrow A+B$$

The concentrations of A, B, C, and D change with time until a state is reached at which both reactions are taking place at the same rate. The concentrations (or pressures) of the components are then constant—the system is said to be in a state of chemical equilibrium. Note that the equilibrium is a dynamic one ; the reactions still take place but at equal rates. The relative proportions of the components determine the 'position' of the equilibrium, which may be *displaced* by changing the conditions (*e.g.*, temperature or pressure).

Equilibrium Constant. According to the law of mass action, for any reversible chemical reaction $aA+bB \rightleftharpoons cC+dD$, the equilibrium constant (K) is defined as

$$K=\frac{[C]^c[D]^d}{[A]^a[B]^b},$$

where $[A]^a$..., etc. are the active masses of A..., etc., more accurately termed the thermodynamic activities.

In thermodynamic terms the equilibrium constant is related to the standard chemical potential by the equation

$$\Delta\mu^\circ=-RT \ln K,$$

where R is the gas constant.

For gas reactions, pressures are often used instead of concentration. The equilibrium constant is then K_p, where $K_p=K_c^n$. Here n is the number of moles of product minus the number of moles of reactant ; for instance, in

$$3H_2+N_2=2NH_3$$

n is $2-(1+3)=-2$.

For gas reactions where the gases are assumed to follow ideal behaviour this equation becomes $\Delta G^\circ=-RT \ln K_p$, where K_p is defined in terms of the partial pressures of reactants and products. Thus for the general reaction above,

$$K_p = \left(\frac{pC^c : pD^d}{pA^a . pB^b} \right).$$

Equilibrium Diagram. A simplified boiling point diagram which shows graphically, for a liquid mixture, the composition of the vapour or solid which is in equilibrium with the liquid. Widely used in metallography to show the equilibrium between liquid and solid alloys.

Equilibrium Law. See equilibrium constant.

Equilibrium, Metastable. If a system, capable thermodynamically of undergoing a spontaneous change, persists without change, it is said to be in a state of metastable equilibrium. The change requires initiation. *e.g.*, a mixture of hydrogen and oxygen at 298 K is capable of undergoing spontaneous reaction but reaction only occurs when initiated by a spark or addition of a catalyst.

Equipartition of Energy. The principle that the total energy of a molecule is, on average, equally distributed among the available degrees of freedom. It is only approximately true in most cases. See degrees of freedom.

Eutectic. Refers to a mixture of two substances in such proportions that no other mixture of the same substances has a lower freezing point.

If a solution containing the proportions of the eutectic is cooled no solid is deposited until the freezing point of the eutectic is reached, when two solid phases are simultaneously deposited in the same proportions as the mixture ; these two solid phases form the eutectic. If one of the substances is water the eutectic can also be referred to as a *cryohydrate*.

Eutectic Mixture. A solid solution consisting of two or more substances and having the lowest freezing point of any possible mixture of these components. The minimum freezing point for a set of components is called the *eutectic point*. Low-melting-point alloys are usually eutectic mixtures.

Eutectic Point. The point on the phase diagram of a mixture of substances which represents the melting point and the composition of the eutectic mixture. The presence of eutectic points in mixtures of substances may be taken to indicate the formation of compounds between the components of the mixture in many, although by no means all, cases.

Evaporation. The change of state of a liquid into a vapour at a temperature below the boiling point of the liquid. Evaporation occurs at the surface of a liquid, some of those molecules with the highest kinetic energies escaping into the gas phase. The result is a fall in the average kinetic energy of the molecules of the liquid and consequently a fall in its temperature.

All liquids and solids have a characteristic vapour pressure depending on the temperature. In a closed vessel, after a sufficient time has elapsed for equilibrium to be established, as many molecules leave the liquid surface to form vapour as return to it from the vapour phase to form liquid. In an open vessel, however, no such equilibrium is set up and if the molecules of vapour immediately over the liquid surface are removed, further liquid molecules vaporize to take their place. In this way the liquid bulk is continuously diminished and the process of evaporation occurs. In the chemical and process industries the term normally refers to the removal of water from a solution.

Exchange Reaction. A reaction which occurs without chemical change. Usually applied to reactions involving isotopic replacement, *e.g.*,

$$^{16}O_2+2H_2{}^{18}O \rightleftharpoons {}^{18}O_2+2H_2{}^{16}O$$

or $H_2+D_2 \rightleftharpoons 2HD$. Because of differences in zero point energies caused by the differing masses of the isotopes, the equilibrium constants in such reactions do not have a value of unity. Use of this fact is made in exchange reactions for the separation of isotopes.

Exciplex. A molecular charge transfer complex formed by the complexing of a donor (or acceptor) in an excited state with an acceptor (or donor) in its ground state. An exciplex is only stable in an energetically excited state.

Excited State. A state of an atom, molecule, etc., when the species has absorbed energy and become excited to a higher energy state as compared with the ground state. The excitation may be electronic, vibrational, rotational, etc.

Exclusion Principle. In any atom no two electrons may have the same set of all four quantum numbers.

Exothermic. Denoting a chemical reaction that releases heat into its surroundings. *Compare* endothermic.

Extinction Coefficient (α). A quantity which is characteristic of a medium which absorbs light. The extinction coefficient (α) is defined as the reciprocal of the layer thickness, measured in centimetres, in which the intensity of radiation is reduced to $\frac{1}{10}$ of its original value. The extinction and absorption coefficients (K) are related ; $\alpha=0.4343$ K. See Beer's law.

F

Face-Centred Cubic. The cubic close-packed lattice (ccp). See face-centred lattice.

Face-Centred Lattice. A lattice having an atom at the face-centre of each unit cell.

Fahrenheit Scale. A temperature scale in which (by modern definition) the temperature of boiling water is taken as 212 degrees and the temperature of melting ice as 32 degrees. It was invented in 1714 by the German scientist G.D. Fahrenheit (1686-1736), who set the zero at the lowest temperature he knew how to obtain in the laboratory (by mixing ice and common salt) and took his own body temperature as 96°F. The scale is no longer in scientific use. To convert to the Celsius scale the formula is $C=5(F-32)/9$.

Farad. Symbol F. The SI unit of capacitance, being the capacitance of a capacitor that, if charged with one coulomb, has a potential difference of one volt between its plates. 1 F=1 CV^{-1}. 1 farad=1 coulomb $volt^{-1}$. The farad itself is too large for most applications ; the practical unit is the microfarad (10^{-6} F). The unit is named after Michael Faraday (1791-1867).

Faraday. Symbol : F. A unit of electric charge equal to the charge required to discharge one mole of a singly charged ion. One faraday is $9.648\ 670 \times 10^4$ coulombs. See Faraday's laws.

Faraday Constant. Symbol : F. The electric charge carried by one mole of electrons or singly ionized ions, *i.e.*, the product of the Avogadro constant and the charge on an electron (disregarding sign). It has the value $9.648\ 670 \times 10^4$ coulombs per mole. This number of coulombs is sometimes treated as a unit of electric charge called the *faraday*.

Faraday Effect. A plane polarized light wave may be considered to consist of an electrical component and a magnetic component which are mutually perpendicular. When such a wave of plane polarized light is passed through a solid or liquid, which is placed in a magnetic field, the plane of polarization is rotated. The phenomenon is called the Faraday effect. Positive rotations are observed when the direction of rotation is the same as that of the magnetizing current.

Faraday's Laws (of electrolysis). Two laws resulting from the work of Michael Faraday on electrolysis :

(1) The amount of chemical change produced is proportional to the charge passed.

(2) The amount of chemical change produced by a given charge depends on the ion concerned. More strictly it is proportional to the relative ionic mass of the ion, and the charge on it. Thus the charge Q needed to deposit *m* grams of an ion of relative ionic mass, M, carrying a charge Z is given by :

$$Q = FmZ/M$$

F, the Faraday constant, has a value of one faraday, *i.e.*, $9.648\ 670 \times 10^4$ coulombs.

These are the modern forms of the laws. Originally, they were stated by Faraday in a different form :

(1) The amount of chemical change produced is proportional to the quantity of electricity passed.

(2) The amount of chemical change produced in different substances by a fixed quantity of electricity is proportional to the electrochemical equivalent of the substance.

F.C.C. Face-centred cubic. See cubic crystal.

F Centre. An anionic site in a crystal occupied only by an electron, *e.g.*, NaCl plus Na vapour gives blue Na_{1+x} Cl containing F centres.

Ferroelectrics. Compounds which show dielectric hysteresis, *i e.*, a reversible spontaneous polarization due to non-cancellation of the elementary dipoles in a crystal. Examples include the perovskite $BaTiO_3$ and KH_2PO_4.

Femto. Symbol : *f.* A prefix used in the metric system to denote 10^{-15}. For example, 10^{-15} second = 1 femtosecond (*fs*).

Fermi. A unit of length formerly used in nuclear physics. It is equal to 10^{-15} metre. In SI units this is equal to 1 femtometre (fm). It was named after the Italian-born US physicist Enrico Fermi (1901-54).

Fermi-Dirac Statistics. See quantum statistics.

Fermi Level. The energy level in a solid at which the probability of finding an electron is 1/2. The Fermi level in conductors lies in the conduction band (see energy bands), in insulators it lies in the valence band, and in semiconductors it falls in the gap between the conduction band and the valence band. At absolute zero all the electrons would occupy energy levels up to the Fermi level and no higher levels would be occupied.

Ferroelectric Materials. Ceramic dielectrics, such as Rochelle salt and barium titanate, that have a domain structure making them analogous to ferromagnetic materials. They exhibit hysteresis and usually the piezoelectric effect.

Fick's Law of Diffusion. A law which relates the rate of diffusion of a substance in a given direction to the gradient of its concentration.

Thus

$$N_A = -D_{AB}\frac{dC_A}{d_z}$$

where N_A is the molar rate of diffusion/unit area ; D_{AB} is the diffusivity of substance A in substance B ; C_A is the molar concentration of A ; Z is the distance in the direction of diffusion.

Fine Structure. Closely spaced lines seen at high resolution in a spectral line or band. Fine structure may be caused by vibration of the molecules or by electron spin. *Hyperfine structure,* seen at very high resolution, is caused by the atomic nucleus affecting the possible energy levels of the atom.

First-Order Reaction. A reaction in which the rate of reaction is proportional to the concentration of one of the reacting substances. The concentration of the reacting substance is raised to the power one ; *i.e.*, rate=k[A]. For example, the decomposition of hydrogen peroxide is a first-order reaction,

$$\text{rate}=k[H_2O_2]$$

Similarly the rate of decay of radioactive material is a first order reaction,

$$\text{rate}=k[\text{radioactive material}]$$

For a first-order reaction, the time for a definite fraction of the reactant to be consumed is independent of the original concentration. The units of k, the rate constant, are s^{-1}.

Fission Products. Refer to the products of the fission of the heavy nuclei. The mass numbers of the fission products range from 30 to 160. Fission is generally an asymmetric process, maximum yields occurring for products of mass ~90 and mass ~140, the distribution of products depending upon the energy of the bombarding particle. Fission products are generally intensely radioactive, many of them decay by β-emission to stable isobars. The separation and disposal of these intensely active materials is a major problem in nuclear energy processes.

Fixed Point. Refers to a temperature that can be accurately reproduced to enable it to be used as the basis of a temperature scale.

Flash Photolysis. A technique which is used for investigating free radicals in gases. The gas is held at low pressure in a long glass or quartz tube, and an absorption spectrum taken using a beam of light passing down the tube. The gas can be subjected to a very brief intense flash of light from a lamp outside the tube, producing free radicals, which are identified by their spectra. Measurements of the intensity of spectral lines can be made with time using an oscilloscope, and the kinetics of very fast reactions can thus be investigated.

Fluidization. A technique which is used in some industrial processes in which solid particles suspended in a stream of gas are treated as if they were in the liquid state. Fluidization is useful for transporting powders, like coal dust. *Fluidized beds*, in which solid particles are suspended in an upward stream, are extensively used in the chemical industry, particularly in catalytic reactions where the powdered catalyst has a high surface area.

Fluorescence. The absorption of energy by atoms, molecules, etc., followed by immediate emission of electromagnetic radiation as the particles make transitions to lower energy states.

Many substances absorb electromagnetic radiation, *e.g.*, light or X-rays, of a certain wavelength. Part of the energy thus absorbed is re-emitted as radiation of longer wavalength. This process is known as fluorescence. The absorption of the original radiation causes an electron to be promoted to a higher energy level in a molecule or atom of the absorbing substance. This excited electron then returns to its original level in a series of steps, each corresponding to an intermediate energy level. X-ray fluorescence is extensively used in the analysis of geological rock specimens. Fluorescence is used as a conventional analytical technique.

Fluorimetry. An analytical technique using the fluorescent properties of the substance to be estimated.

Fluorite Structure. Calcium-fluoride structure.

Fluxional. Molecules which rearrange so easily on a particular time scale (*e.g.* n.m.r.) that the rearrangement process can be detected are termed fluxional or stereochemically non-rigid. Example, PF_5 which although it is a trigonal bipyramid shows only one type of fluorine by ^{19}F n.m.r. spectroscopy. The rearrangements can often be shown/and studied by reduction of temperature. Molecules which are fluxional over all available temperatures are termed fictile.

Foam. Refers to a dispersion of bubbles in a liquid. Foams can be stabilized by surfactants. Solid foams (*e.g.*, expanded polystyrene or foam rubber) are made by foaming the liquid and allowing it to set.

Force Constant. A measure of the restoring force to vibration in a bond. It is given by

$$v = \frac{1}{c}\sqrt{\frac{f}{4\pi^2\mu}}$$

where v is the vibrational frequency, (cm^{-1}), f is the force constant, c is the velocity of light and μ is the reduced mass of the oscillator A—B(g). The stronger the bond the greater the force constant.

Fortin Barometer. See barometer.

Fraction. A mixture of liquids with similar boiling points collected by fractional distillation.

Fractional Distillation (fractionation). A distillation carried out with partial reflux, using a long vertical column (fractionating column). It utilizes the fact that the vapour phase above a liquid mixture is generally richer in the more volatile component. If the region in which refluxing occurs is sufficiently long, fractionation permits the complete separation of two or more volatile liquids. Fractionation is the fundamental process for producing petroleum from crude oil.

Unlike normal reflux, the fractionating column may be insulated to reduce heat loss, and special designs are used to maximize the liquid-vapour interface.

The simplest method of doing fractional distillation is by gradually vaporizing the liquid mixture, and removing and condensing the vapour as it is evolved. During the process the b.p. of the liquid gradually rises, due to loss of more volatile components, and the condensate may be collected in a series of fractions of different boiling ranges. The disadvantage of this method, which is known as ***differential distillation***, is that since the vapour evolved at any time is in equilibrium with the boiling liquid, unless there are marked differences in the boiling points of the components separation will be poor. Consequently this method is rarely used outside the laboratory.

Rectification is much more efficient, and is the fractionation process normally employed. Fractional distillation may be carried out at low temperatures, generally under vacuum.

Fractionating Column. See rectification.

Free Energy. A measure of the ability of a system to do useful work. It is a thermodynamic state function which is represented by G (after Willard Gibbs). The free energy change ($\triangle G$) in any system is related to the enthalpy and entropy changes by the equation :

$$\triangle G = \triangle H - T\triangle S.$$

$\triangle G$ is a measure of the maximum amount of *useful* work which may be obtained from the change under consideration. In any system the value of the free energy change determines the position of equilibrium in that at equilibrium the free energy is a minimum, *i.e.*, $\triangle G=0$ at equilibrium. For substances in their standard states, *i.e.*, 1 mole at 1 atmosphere pressure and a specified temperature, the change in free energy is called the *standard free energy change*, $\triangle G_T°$. This latter quantity is of considerable importance in that it determines the thermodynamic feasibility of a reaction. For a reaction to be thermodynamically feasible and thus for more products to be formed than reactants at equilibrium, the value of $\triangle G°$ must be negative, *i.e.*, there must be a decrease in standard free energy from reactants to products. $\triangle G$ and $\triangle G°$ are usually expressed in kilojoules per mole for chemical reactions.

The *Helmholtz free energy* (or *Helmholtz function*), F, is defined by $F=E-TS$, where E is the internal energy. For a reversible isothermal process, $\triangle F$ represents the useful work available.

Free Radical. An atom or group of atoms with an unpaired valence electron. Free radicals can be produced by photolysis or pyrolysis in which a bond is broken without forming ions (see homolytic fission). Because of their free valency, most free radicals are extremely reactive.

Free radicals are produced by breaking a covalent bond ; for example :

$$CH_3Cl \rightarrow CH_3\cdot + Cl.$$

They are often formed in light-induced reactions. Free radicals are extremely reactive and can only be stabilized and isolated under special conditions.

Freezing. The process by which a liquid is converted into a solid by cooling ; the reverse of melting.

Freezing Mixture. A mixture of components that produces a low temperature. For example, a mixture of ice and sodium chloride gives a temperature of $-20°C$.

Freezing Point. The temperature at which a liquid is in equilibrium with its solid phase at standard pressure and below which the liquid freezes or solidifies. This temperature is always the same for a particular liquid and is numerically equal to the melting point of the solid.

Freezing Point Depression. Addition of a solute to a solvent causing a depression of the freezing point of the latter by an amount which is proportional to the number of solute particles in the solution. The depression caused by dissolving 1 mole of the solute in 1 litre of solvent is termed the molecular depression or cryoscopic constant of the solvent. See colligative properties.

Frequency, v. The number of wavelengths passing a fixed observer per second. Thus v depends on the velocity of the wave motion and upon its wavelength. A frequency (wave number) used in spectroscopy is the number of complete waves per cm (cm^{-1}), the reciprocal of the wavelength. Frequency is proportional to energy.

Freundlich Isotherm. The relationship between the amount of a substance adsorbed and the concentration of the solute

$$\frac{x}{m}=kc^{\frac{1}{n}}$$

where x is the amount absorbed, m the weight of adsorbent, c the concentration or pressure, and k and n constants. This equation is empirical, depending on no theoretical assumptions. The values of the constants k and n vary with the system ; n is usually between 0.2 and 0.7. The Freundlich, equation is successful with adsorption systems of very variable characteristics, such as adsorption from solution and adsorption of gases. It expresses the rapid increase of adsorption with concentration at low concentrations, but is not so successful at higher concentrations.

Fuel Cell. A type of cell in which fuel is converted directly into electricity. Large fuel cells can generate tens of amperes. Usually the e.m.f. is about 0.9 volt and the efficiency around 60%.

The simplest fuel cell is one in which hydrogen is oxidized to form water over porous sintered nickel electrodes. A supply of gaseous hydrogen is fed to a compartment containing the porous cathode and a supply of oxygen is fed to a compartment containing the porous anode ; the electrodes are separated by a third compartment containing a hot alkaline electrolyte, such as potassium hydroxide. The electrodes are porous to enable the gases to react with the electrolyte, with the nickel in the electrodes acting as a catalyst. At the cathode the hydrogen reacts with the hydroxide ions in the electrolyte to form water, with the release of two electrons per hydrogen molecule :

$$H_2 + 2OH^- \rightarrow 2H_2O + 2e^-$$

At the anode, the oxygen reacts with the water, taking up electrons, to form hydroxide ions :

$$\tfrac{1}{2}O_2 + H_2O + 2e^- \rightarrow 2OH^-$$

The electrons flow from the cathode to the anode through an external circuit as an electric current. The device is a more effcient converter of electric energy than a heat engine, but it is bulky and requires a continuous supply of gaseous fuels use to power electric vehicles is being actively explored.

Fuels which have been used include hydrogen, hydrazine, methanol and ammonia, while oxidants are usually oxygen or air. Electrolytes comprise alkali solutions, molten carbonates, solid oxides, ion-exchange resins, etc.

Batteries of fuel cells have been used in the American aerospace programme but, in general, the potentially high efficiency of the fuel cell has not been realized industrially.

Fugacity. Symbol *f*. A thermodynamic function used in place of partial pressure in reactions involving real gases and mixtures. For a component of a mixture, it is defined by $d\mu = RTd(\ln f)$, where μ is the chemical potential. It has the same units as pressure and the fugacity of a gas is equal to the pressure if the gas is ideal. The fugacity of a liquid or solid is the fugacity of the vapour with which it is in equilibrium. The ratio of the fugacity to the fugacity in some standard state is the activity. For a gas, the standard state is chosen to be the state at which the fugacity is 1. The activity then equals the fugacity.

Fundamental Constants (Universal Constants). Those parameters that do not change throughout the universe. The charge on an electron, the speed of light in free space, the Planck constant, the gravitational constant, the electric constant, and the magnetic constant are all thought to be examples.

Fundamental Units. The set of units of length, mass, and time that form the basis of most systems of units.

Such a set requires three mechanical units (usually of length, mass, and time) and one electrical unit ; it has also been found convenient to treat certain other quantities as fundamental, even though they are not strictly independent. In the metric system the centimetre—gram—second (c.g.s.) system was replaced by the metre—kilogram—second (m.k.s.) system ; the latter has now been adapted to provide the basis for SI units. In British Imperial units the food—pound—second (f.p.s.) system was formerly used.

Fusion. Melting.

G

Galvanized Iron. Iron or steel that is coated with a layer of zinc to protect it from corrosion. Corrugated mild-steel sheets for roofing and mild-steel sheets for dust-bins, etc., are generally galvanized by dipping them in molten zinc. The formation of a brittle zinc-iron alloy is prevented by the addition of small quantities of aluminium or magnesium. Wire is usually galvanized by a cold electrolytic process as no alloy forms in this process. Galvanizing has been an effective method of protecting steel because even if the surface is scratched, the zinc still protects the underlying metal.

Galvanic Cell. See cell.

Galvanizing. The coating of steel with Zn for protection by dipping into molten Zn or by electrodeposition (cold galvanizing). Corrosion takes place preferentially at the Zn.

Gamma Radiation. Electromagnetic radiation which is emitted by excited atomic nuclei during the process of passing to a lower excitation state. Gamma radiation ranges in energy from about 10^{-15} to 10^{-11} joule (10 keV to 100 MeV) corresponding to a wavelength range of about 10^{-10} to 10^{-14} metre. A common source of gamma radiation is cobalt-60, the decay process of which is :

$$^{60}_{27}Co \rightarrow \ ^{60}_{28}Ni \rightarrow Ni$$

The de-excitation of nickel-60 gets accompanied by the emission of a gamma-ray photon having an energy of 1.87×10^{-13} J (1.17 MeV).

Gas. A state of matter in which the matter concerned occupies the whole of its container irrespective of its quantity. In an ideal gas, which obeys the gas laws exactly, the molecules themselves would have a negligible volume and negligible forces between them, and collisions between molecules would be perfectly elastic. In practice, however, the behaviour of real gases deviates from the gas laws because their molecules occupy a finite volume, there are small forces between molecules, and in polyatomic gases collisions are to a certain extent inelastic.

The gaseous state is the most diffuse state of matter in which molecules have almost unrestricted motion. A substance for which the volume increases continuously and without limit as the pressure is continuously reduced. A vapour is distinguished from a gas in that a gas is above the critical temperature of the substance.

The particles of gas have freedom of movement and gases, therefore, have no fixed shape or volume. The atoms and molecules of a gas are in a continual state of motion and are continually colliding with each other and with the walls of the containing vessel. These collisions with the walls create the pressure of a gas.

Gas Absorption. The solution of one component of a gaseous mixture in a liquid or the absorption on to a surface. Selective absorption is used to separate components of mixtures of gases or vapours, *e.g.*, the removal of NH_3 from NH_3—air mixtures by absorption in water, or the removal of benzene from coal gas by washing with absorbent oils.

On an industrial scale the process may be carried out by bubbling the gas through the liquid (as in a *plate absorption column*), by spraying the liquid through the gas (as in a *spray column*) or by passing the gas over surfaces wetted by the liquid (as in a packed *absorption tower*). Counter-current flow is generally employed.

Gas Calorimeter. Equipment for the determination of the calorific value of a fuel gas. In flow calorimeters gas at constant pressure is allowed to burn at a known rate in a burner and the heat evolved is absorbed by water flowing through the calorimeter. The calorific value can be calculated from the volume of gas burned and the temperature rise of a known mass of water. The best known flow calorimeters are the Boys and the Junkers types.

For small amounts of combustible gas a still-water calorimeter is used. This involves combustion or explosion of a known volume of gas, the temperature rise of a surrounding mass of water being measured.

Gas Constant. By combining Boyle's and Charles's laws, we arrive at a general equation of the form $pv=k\mathrm{T}$, where p is the pressure, v the volume and T the temperature of the gas, while k is a constant depending on the amount of gas used, and the units employed for expressing p and v. When 1 mole of gas is considered, k may be replaced by R, and the equation becomes $pv=\mathrm{RT}$. R is called the gas constant, R=0,08204 litre-atmos. per degree, or $8.31434/(\mathrm{mol})^4 \times 10^7$ ergs per degree, or 1.987 cal per mole per degree or 8.31434 $\mathrm{JK}^{-1}\ \mathrm{mol}^{-1}$.

Gas Constant (Universal Molar Gas Constant). Symbol R. The constant that appears in the *universal gas equation* (see gas laws). It has the value 8.31434 J K^{-1} mol^{-1}.

Gaseous Diffusion Separation. A technique which is used for the separation of gases that relies on their slightly different atomic masses. For example, the isotopes of uranium (^{235}U and ^{238}U) can be separated by first preparing uranium hexafluoride from uranium ore. If this is then made to pass through a series of fine pores, the molecules containing the lighter isotope of uranium will pass through more slowly. As the gases pass through successive 'diaphragms', the proportion of lighter molecules increases and a separation of over 99% is attainable. This method is also applied to the separation of the isotopes of hydrogen.

Gas Equation. See gas laws.

Gas Laws. Laws which are relating the temperature, pressure, and volume of a fixed mass of gas. The main gas laws are Boyle's law and Charles' law. The laws have been not obeyed exactly by any real gas, but many common gases obey them under certain conditions, particularly at high temperatures and low pressures. A gas that would obey the laws over all pressures and temperatures is a *perfect* or *ideal gas.*

Boyle's and Charles' laws can be combined into an equation of state for ideal gases :

$$pV_m = RT$$

where V_m refers to the molar volume and R the molar gas constant. For *n* moles of gas

$$pV = nRT$$

All real gases deviate to some extent from the gas laws, which are applicable only to idealized systems of particles of negligible volume with no intermolecular forces. There have been several modified equations of state that give a better description of the behaviour of real gases, the best known being the van der Waals equation.

The gas laws were first established experimentally for real gases, although they are obeyed by real gases to only a limited extent ; they are obeyed best at high temperatures and low pressures. See *also* equation of state.

Gay-Lussac's Law. The law was first stated in 1808 by J.L. Gay Lussac (1778-1850) and led to Avogadro's law. When gases combine they do so in volumes which are in a simple ratio to each other, and to that of the product if it is also gaseous. Thus 1 vol of nitrogen combines with 3 vol of hydrogen to form 2 vol of ammonia. The law is only approximately true and only strictly holds for ideal gases.

Charles's law, which is a quite different law, is also sometimes referred to as Gay-Lussac's law.

Gas Thermometer. A device which is used for measuring temperature in which the working fluid is a gas. It provides the most accurate method of measuring temperatures in the range 2.5 to 1337 K. Using a fixed mass of gas a *constant-volume thermometer* is able to measure the pressure of a fixed volume of gas at relevant temperatures, usually by using a mercury manometer and a barometer.

Gauss. Symbol : G. The unit of magnetic flux density in the c.g.s. system. It is equal to 10^{-4} tesla.

Gel. Refers to a lyophilic colloid that is normally stable but may be induced to coagulate partially under certain conditions (*e.g.*, lowering the temperature). This produces a pseudo-solid or easily deformable jelly-like mass, called a gel, in which intertwining particles enclose the whole dispersing medium. Gels may be further subdivided into elastic gels (*e.g.*, gelatin) and rigid gels (*e.g.*, silica gel).

Hydrophilic colloids are capable under certain conditions, such as lowering of temperature, of partially coagulating to a mass of intertwining filaments which may enclose the whole of the dispersion medium to produce a pseudo-solid but easily deformable mass or jelly. Such gels may, in some instances, preserve a rigidity when they contain as little as 1% of disperse phase, and generally appear to be heterogeneous under the microscope. Gels are sometimes classified as hydrogels, alcogels, etc., according to whether the dispersion medium is water, alcohol, etc.

Inorganic gelatinous precipitates have much the same type of structure as coherent gels, and can often be made to form true gels on careful choice of conditions. Gels are sometimes divided into elastic and rigid gels. Gelatin is an example of the first type, silica gel of the second, but this classification is not strict, as even silica has a certain amount of elasticity. Gelatin gel may be converted back to the sol on heating and is thus a reversible gel ; silica will not liquefy to a sol on any simple treatment, and is said to be irreversible. See colloids.

Gelatin(e). Refers to a colourless or pale yellow water-soluble protein which is obtained by boiling collagen with water and evaporating the solution. It swells when water is added and dissolves in hot water to form a solution that sets to a gel on cooling. It finds use in photographic emulsions and adhesives, and in jellies and other foodstuffs.

Gel Filtration. Refers to a type of column chromatography in which a mixture of liquids is passed down a column containing a gel. Small molecules in the mixture can enter pores in the gel and move slowly down the column ; large molecules, which cannot enter the pores, move more quickly. Thus, mixtures of molecules can be separated on the basis of their size. The technique finds use particularly for separating proteins but it can also be applied to other polymers and to cell nuclei, viruses, etc.

Gibb's Equation of Surface Concentration. This equation relates the surface tension (γ) of a solution and the amount (λ) of the solute adsorbed at unit area of the surface. For a single non-ionic solute in dilute solution the equation approximates to

$$\lambda = -\frac{c}{RT}\frac{d\gamma}{dc}$$

(where c is the concentration per unit volume and R is the gas constant).

Qualitatively the equation shows that solutes which lower the surface tension have a positive surface concentration, *e.g.*, soaps in water or amyl alcohol in water. Conversely solutes which increase the surface tension have a negative surface concentration.

Gibbs Free Energy (Gibbs Function). See Gibbs function. It is named after the US chemist J.W. Gibbs (1839-1903).

Gibbs Function (Gibbs Free Energy). Symbol : G. A thermodynamic function defined by

$$G=H-TS$$

where H is the enthalpy, T the thermodynamic temperature, and S the entropy. It is useful for specifying the conditions of chemical equilibrium for reactions for constant temperature and pressure (G is a minimum). See *also* free energy.

Gibbs-Helmholtz Equation. This equation relates the heats and free energy changes which occur during a chemical reaction. For a reaction carried out at constant pressure

$$\Delta G=\Delta H-T\Delta S$$

where ΔG is the Gibbs free energy change, ΔH is the enthalpy change and ΔS, the entropy change. For a reaction carried out at constant volume the corresponding equation is

$$\Delta A=\Delta E+T\left(\frac{dA}{dT}\right)_v=\Delta E+T\Delta S$$

where ΔA is the change in Helmholtz free energy and ΔE is the change in internal energy.

Using this equation it is possible to calculate heats of reaction from the variation of ΔG with temperature.

Giga. Symbol G. A prefix which is used in the metric system to denote one thousand million times. For example, 10^9 joules= 1 gigajoule (GJ).

Glass Electrode. A type of half cell having a glass bulb containing an acidic solution of fixed pH, into which dips a platinum wire. The glass bulb is thin enough for hydrogen ions to diffuse through. If the bulb is kept in a solution having hydrogen ions, the electrode potential depends on the hydrogen-ion concentration. Glass electrodes find use in pH measurement.

Globulin. The term used for any of a group of globular proteins that are generally insoluble in water and present in blood, eggs, milk, and as a reserve protein in seeds. Blood serum globulins comprise four types : α_1-, α_2-, and β-globulins, which serve as carrier proteins ; and γ-globulins, which include the immunoglobulins responsible for immune responses.

Glow Discharge. When an electrical potential is applied to a gas at low pressure, the gas acts as a conductor, and becomes luminous. The appearance of the discharge is having certain characteristic features. Thus at about 0.1 mm in helium there is a narrow dark space next to the cathode (the Aston dark space), with a thin glowing layer of gas next to it (the cathode glow), which is followed by a wider dark space (the Crookes or cathode dark space), then a glowing area (the negative glow), which is separated by the Faraday dark space from a longer area of glowing gas (the positive column), which stretches as far as the anode. The exact dimensions of the various dark spaces and glows vary with the potential, the gas pressure and the nature of the gas. At pressures even lower than 0.1 mm in helium, the discharge consists only of a negative glow and the Crookes' dark space.

Gold Number. Refers to a number which is used to define the efficiency of operation of protective colloids based upon the amount of protective colloid preventing the colour change of a gold sol from red to blue on coagulation by elecrolytes. The smaller the gold number the more efficient has been the protective colloid. The gold numbers of a few protective colloids are gelatine 0.005-0.01, sodium caseinate 0.01, haemoglobin 0.03-0.07, albumin 0.1-0.2, starch 25.

Gouy Balance. A balance which is used for the determination of magnetic susceptibility. The sample is weighed in and out of a magnetic field and the susceptibility is calculated from the difference in weights.

Graham's Law (of Diffusion). The principle that gases diffuse at a rate that is inversely proportional to the square root of their density. Light molecules diffuse faster than heavy molecules.

$$\left(\frac{\text{rate A}}{\text{rate B}}\right)=\sqrt{\frac{\rho B}{\rho A}}$$

The principle is used in the separation of isotopes.

The law was formulated in 1829 by Thomas Graham (1805-69).

Gram. Symbol *g.* One thousandth of a kilogram. The gram is the fundamental unit of mass in c.g.s. units and was formerly used in such units as the *gram-atom, gram-molecule,* and *gram-equivalent,* which have now been replaced by the mole.

Gram Atom. The quantity of an element numerically equal to the atomic weight expressed in grams.

Gram Equivalent. The equivalent weight of a substance expressed in grams.

Gram-Equivalent. The equivalent weight of a substance in grams.

Gram Molecular Volume. The volume occupied by the gram molecule of an element or compound in the gaseous state. According to Avagadro's hypothesis, under the same conditions of temperature and pressure, all gases have the same gram molecular volume. At s.t.p. the gram molecular volume is equal to 22.414 litres.

Gram Molecule. The quantity of a compound, or element, equal numerically to the molecular weight expressed in grams. The weight in grams of one mole of a substance.

Graphite. An allotrope of carbon. Graphite is a good conductor of heat and electricity. The atoms are arranged in layers. Graphite finds use as a solid lubricant.

Gray. Symbol Gy. The derived SI unit of absorbed dose of ionizing radiation (see radiation units). It is named after the British radiobiologist L.H. Gray (1905-65).

Grottius-Draper Law. This law states that only light which is absorbed by a substance is effective in inducing a chemical change. It is not correct to consider that all the light absorbed brings about reaction, since some of it may be re-emitted as fluorescence or as heat. It is not necessary to absorb the light directly by the reacting substances ; it is possible, as in photosensitization, to absorb the light by an inert substance which then transfers the stored energy to the reactants as thermal energy.

Ground State. 1. The lowest energy electronic, vibrational, or rotational state of an atom, molecule or ion.

2. The lowest energy state of an atom, molecule, or other system. *Compare* excited state.

Group Theory. A mathematical method which considers the effect of a group of operators (*e.g.*, symmetry elements, crystal field) on properties. Used in calculations on structure, spectra, magnetic susceptibility, etc.

Gyromagnetic Ratio. For a nucleus the gyromagnetic ratio is equal to the ratio of the magnitude of the magnetic moment (μ) and the angular momentum (I), *i.e.*,

$$\mu = gI \quad \text{where} \quad I = h\sqrt{I(I+1)},$$

I being the nuclear spin quantum number.

H

Habit. See crystal habit.

Half Cell. Refers to an electrode which is in contact with a solution of ions. In general there will be an e.m.f. set up between electrode and solution by transfer of electrons to or from the electrode. The e.m.f. of a half cell cannot be measured directly since setting up a circuit results in the formation of another half cell. See electrode potential

Half-life, Half-value Period, $t\frac{1}{2}$. The time taken for the concentration of a substance to fall to half its initial value. For radioactive elements the half-life is $0.69 \times 1/\lambda$ where λ is the decay constant. The term is used for transient species (*e.g.* free radicals) and first order chemical reactions.

Half Reaction. Most generally applied to redox reactions the half reaction refers to the stoicheiometry of the reaction of one of the species involved, *e.g.*

$$MnO_4^- + 8H + 5e \longrightarrow Mn^{2+} + 4H_2O$$

Half-thickness. The thickness of a specified material that reduces the intensity of a beam of radiation to half its original value.

Half-value Period. See half-life.

Half Wave Potential, $E\frac{1}{2}$. The midpoint of the polarographic wave of an electrochemically reversible reaction. $E\frac{1}{2}$ is characteristic of the species under investigation.

Half-width. Half the width of a spectrum line (or in some cases the full width) measured at half its height.

Hall–Heroult Cell. An electrolytic cell used industrially for the extraction of aluminium from bauxite. The bauxite is first purified by dissolving it in sodium hydroxide and filtering off insoluble constituents. Aluminium hydroxide is then precipitated (by adding CO_2) and this is decomposed by heating to obtain pure Al_2O_3. In the Hall-Heroult cell, the oxide is mixed with cryolite (to lower its melting point) and the molten mixture electrolysed using graphite anodes. The cathode is the lining of the cell, also of graphite. The electrolyte is kept in a molten state (about 850°C) by the current. Molten aluminium collects at the bottom of the cell and can be tapped off. Oxygen forms at the anode, and gradually oxidizes it away. The ceil is named after the US chemist Charles Martin Hall (1863-1914), who discovered the process in 1886, and the French chemist Paul Heroult (1863-1914), who discovered it independently in the same year.

Hall Effect. If a current (I) is passed through a conducting crystal in a direction perpendicular to that of an applied magnetic field (H), the conductor develops a potential (V) between the faces which are mutually perpendicular to both the direction of the current and the magnetic field. This is known as the Hall effect ; the magnitude of the potential difference is given by

$$V = \frac{IH}{Ned}$$

where N is the number of current carriers, *e* is the electronic charge and *d*, the distance of separation of the two faces across which the potential is developed.

Hardness. The resistance of a material to pressure applied to a small area, *i.e.* resistance to crushing, abrasion, indentation or stretching. The hardness of a crystal will depend upon the degree of crystallinity and will also very in different directions in the crystal.

Hardness is assessed on an empirical ten-point scale, diamond (10) is the hardest substance known, talc (1) is soft. Hardness is measured by scratching with substances of known hardness ; hardness of alloys is measured by a series of tests which measure deformation under standard conditions. Brinell, Rockwell, Vieker's, Knoop and Monotron are particular tests.

Heat. Energy transferred as a result of a temperature difference. The term is often loosely used to mean internal energy (*i.e.* the total kinetic and potential energy of the particles. It is common in chemistry to define such quantities as *heat of combustion, heat of neutralization,* etc. These are in fact molar enthalpies for the change, given the symbol $\triangle H_m$. The unit is usually the kilojoule per mole (kJ mol^{-1}). By convention, $\triangle H$ is negative for an exothermic reaction.

Molar enthalpy changes stated for chemical reactions are changes for standard conditions, which are 298 K (25°C) and 101,325 Pa (1 atmosphere). Thus, the standard molar enthalpy of reaction is the enthalpy change for reaction of substances under these conditions producing reactants under the same conditions. The substances involved must be in their normal equilibrium physical states under these conditions (*e.g.* carbon as graphite, water as the liquid, etc.). Note that the measured enthalpy change will not usually be the standard change. In addition, it is common to specify the entity involved. For instance $\triangle H$ (H_2O) is the standard molar enthalpy of formation for one mole of H_2O species.

Heat Capacity (Thermal Capacity). The ratio of the heat supplied to an object or specimen to its consequent rise in temperature. The *specific heat capacity* is the ratio of the heat supplied to unit mass of a substance to its consequent rise in temperature. The *molar heat capacity* is the ratio of the heat supplied to unit amount of a substance to its consequent rise in temperature. In practice, heat capacity (C) is measured in joules per kelvin, specific heat capacity (c) in JK^{-1} kg^{-1}, and molar heat capacity (C_m) in JK^{-1} mol^{-1}. For a gas, the values of c and C_m are commonly given either at *constant volume*, when only its internal energy is increased, or at *constant pressure*, which requires a greater input of heat as the gas is allowed to expand and do work against the surroundings. The symbols for the specific and molar heat capacities at constant volume are c_v and C_v, respectively ; those for the specific and molar heat capacities at constant pressure are c_p and C_p.

Heat Engine. (thermodynamic engine). A device for converting heat energy into work. Heat engines operate by transferring energy from a high temperature source to a low temperature sink. The theoretical operation of heat engines is useful in the theory of thermodynamics.

Heat Exchangers. Devices that enable the heat from a hot fluid to be transferred to a cool fluid without allowing them to come into contact. The normal arrangement is for one of the fluids to flow in a coiled tube through a jacket containing the second fluid. Both the cooling and heating effect may be of benefit in conserving the energy used in a chemical plant and in controlling the process.

Heat of Atomization. The energy required in dissociating one mole of a substance into atoms.

Heat of Combustion. The energy liberated when one mole of a given substance is completely oxidized.

Heat of Crystallization. The heat evolved when unit weight of solute crystallizes from an infinite quantity of a saturated solution.

Heat of Dissociation. The amount of heat required to dissociate 1 mole of a compound into its constituent elements, or into certain specified smaller molecules or other species.

Heat of Formation. The energy change when one mole of a substance is formed from its elements.

Heat of Formation, Standard ($\triangle H_f^0$). The heat evolved or absorbed when 1 mole of a substance is formed from its constituent elements at 1 atm. pressure and a stated temperature, commonly but not necessarily 298 K. By definition the standard heat of formation of an element in its standard state is taken arbitrarily to be zero.

Heat of Neutralization. The energy liberated when one mole of an acid or base is neutralized.

Heat of Reaction. The amount of heat (usually expressed in kilojoules) absorbed or evolved when specified amounts (usually 1 mole of the product) react under constant pressure conditions. For exothermic reactions the convention is adopted that the enthalpy change, the heat of reaction, is negative. Thus, *e.g.*

$$C(\text{solid}) + \tfrac{1}{2}O_2(\text{gas}) \longrightarrow CO(\text{gas})$$
$$\Delta H = -110.88\text{kJ}$$

110.88 kJ of heat are evolved when 1 mole of solid carbon reacts with ½ mole of oxygen gas at constant pressure.

It is essential to specify the physical states of the reactants and products, since there may be additional heat changes associated with changes in state.

Heat of Solution. The energy liberated or absorbed when one mole of a given substance is completely dissolved in a large volume of solvent (strictly, to infinite dilution).

Hecto. Symbol h. A prefix used in the metric system to denote 100 times. For example, 100 coulombs = 1 hectocoulomb (hC).

Heisenberg Uncertainty Principle. See uncertainly principle.

Helmholtz Free Energy. The maximum amount of energy available to do work resulting from changes in a system at constant volume.

Helmholtz Function. (Helmholtz free energy) Symbol : F. A thermodynamic function defined by

$$F = E - TS$$

where E is the internal energy, T the thermodynamic temperature, and S the entropy. It is a measure of the ability of a system to do useful work in an isothermal process.

Hemihedral Forms. Those forms in any crystal system which show the full number of faces required by the symmetry of the system are called holohedral forms. When only half the number of faces found in the holohedral form are present, the form is said to be hemihedral.

Henderson-Hasselbach Equation. A simplified version of the relationships used in calculations on buffer solutions.

$$pH = pK_a + \log \frac{\text{fraction neutralized}}{\text{fraction unneutralized}}$$

Henry. Symbol : H. The SI unit of inductance, equal to the inductance of a closed circuit that has a magnetic flux of one weber per ampere of current in the circuit. 1 H=1 Wb A^{-1}.

Henry's Law. The concentration (C) of a gas in solution is proportional to the partial pressure (p) of that gas in equilibrium with the solution, *i e.* $p=kC$, where k is a proportionality constant. The relationship is similar in form to that for Raoult's law, which deals with ideal solutions. The law is only obeyed provided there is no chemical reaction between the gas and the liquid.

A consequence of Henry's law is that the 'volume solubility' of a gas is independent of pressure.

Heptahydrate. A crystalline hydrated compound containing one molecule of compound per seven molecules of water of crystallization.

Hertz. Symbol Hz. The SI unit of frequency equal to one cycle per second. It is named after Heinrich, Hertz (1857-94), a German physicist.

Hess's Law. Sometimes called the law of constant heat summation, it states that the total heat change accompanying a chemical reaction is independent of the route taken in reactants becoming products. Hess's law is an application of the first law of thermodynamics to chemical reactions.

The law can be used to obtain thermodynamic data that cannot be measured directly. For example, the heat of formation of ethane can be found by considering the reactions :

$$2C(s) + 3H_2(g) + 3\tfrac{1}{2}O_2(g) \rightarrow$$
$$2CO_2(g) + 3H_2O(1)$$

The heat of this reaction is $2\Delta H_c + 3\Delta H_h$ where ΔH_c and ΔH_h are the heats of combustion of carbon and hydrogen respectively, which can be measured. By Hess' law, this is equal to the sum of the energies for two stages :

$$2C(s) + 3H_2(g) \rightarrow C_2H_6(g)$$

(the heat of formation of ethane, $\triangle H_f$) and $C_2H_6(g)+3\frac{1}{2}O_2 \rightarrow 2CO_2(g)+3H_2O(l)$ (the heat of combustion of ethnae, $\triangle H_e$). As $\triangle H_e$ can be measured and as

$$\triangle H_f+\triangle H_e=2\triangle H_c+3\triangle H_h$$

$\triangle H_f$ can be found. Another example is the used of the Born-Haber cyole to obtain lattice energies. The law was first put forward in 1840 by the Russian chemist Germain Hess (1802-50). It is sometimes called the *law of constant heat summation* and is a consequence of the law of conservation of energy.

Heterogeneous. Relating to more than one phase. A heterogeneous mixture, for instance, contains two or more distinct phases.

Heterogeneous Catalysis. A catalysed reaction where the catalyst is in a different phase from the reactants, *e.g.* a solid catalyst in a reaction between liquid or gaseous reactants. Such reactions are generally considered to involve adsorption and reaction of the reactants at the catalyst surface.

Heterogeneous Reaction. A reaction which occurs between substances in different phases, *e.g.* between a gas and a liquid.

Hexagonal Close Packing. See close packing.

Hexagonal Crystal. See crystal system.

Heyrovsky-Ilkovic Equation. The relation between the half wave potential, $E\frac{1}{2}$, for a polarogram, the current, i, for the potential at the dropping mercury electrode, E_{dme}, the number of electrons, n, involved in the electrochemical reaction (id is the diffusion current).

$$E_{dme}=E_{1/2}+\frac{0.059}{n}\log\frac{id-i}{i}$$

Homogenous. Relating to a single phase. A homogeneous mixture, for instance, consists of only one phase.

Homogeneous Catalysis. The process which occurs when the catalyst is in the same phase as the reactants, *e.g.* the acid-catalysed hydrolysis of an ester may be regarded as a homogeneously catalysed reaction. Recently the term has been used more loosely to describe processes where the catalyst and one of the reactants are in the same phase, *e.g.* the hydrogenation of liquid olefins catalysed by solutions of transition metal complexes.

Homogeneous Combustion. See cool flames.

Homogeneous Reaction. Reactions which occur between substances in the same phase, *e.g.* between gases or liquids. A reaction between two solids is not usually regarded as a homogeneous reaction.

Homopolar Crystal. A crystal with all bonds homopolar (covalent), *e.g.* diamond, zinc blende.

Hydrogen Electrode. Refers to a type of half cell based on hydrogen, and assigned zero electrode potential, so that other elements may be compared with it. It is also called the standard hydrogen half cell. The hydrogen is bubbled over a platinum electrode, coated in 'platinum black', in a 1 M acid solution. Hydrogen is adsosbed on the platinum black, which has a high surface area, enabling the equilibrium

$$H(g) \rightarrow H^+(aq) + e^-$$

to be set up. The platinum is inert and has no tendency to form platinum ions in solution.

For solutions in which the concentration of hydrogen ions in equal to unit activity and for a gas pressure of 1 atm. the electrode potential is said to be normal and is arbitrarily assigned a value of zero. Although other electrode potentials are quoted relative to the standard hydrogen electrode, the latter is seldom used in practice for this purpose. Usually, electrode potentials are measured against a standard calomel electrode which is then referred to the 'hydrogen scale'.

Hydrogen Half Cell. See hydrogen electrode.

Hydrogen-ion Concentration. Hydrogen-ion concentrations are measured as the number of gram ions of hydrogen ions present per litre of solution. Since these concentrations are usually small, the concentration is generally expressed as the pH of the solution ; the pH being the logarithm of the reciprocal of the hydrogen-ion concentration, *i.e.* $pH = -\log[H^+]$. The pH can be measured using a glass electrode or, less accurately, by coloured indicators.

Hydrogen Ions. Chemically free hydrogen ions are rarely formed and dissociation to H^+ occurs only when the resultant ion can be complexed. Thus HCl gas is undissociated but gives $[H(H_2O)_n]^+Cl^-$ and $[NH_4]^+Cl^-$ with H_2O or NH_3.

Hydrogenolysis. The cleaving of a chemical bond by hydrogen, usually in the presence of a hydrogenation catalyst.

Hydrogen Overvoltage. In electrolytic processes where hydrogen is evolved at the cathode, the potential of the latter is usually more negative than that formally required for hydrogen liberation. The cathode, when thus behaving as an irreversible electrode, is said to exhibit an overpotential or overvoltage. The potential differences between the irreversible cathode and the reversible hydrogen electrode in the same solution is the hydrogen overvoltage at the particular cathode.

Hydrogen overvoltage depends upon the electrode (its material, nature of surface, etc.) and increases logarithmically with current density. Since the hydrogen overvoltage has a negative sign, the evolution potential of hydrogen is moved up the electrochemical series, thus enabling the electrodeposition of such metals as Pb, Ni, Zn from acid solutions on to cathodes of these metals.

Hydrolysis. A chemical reaction of a compound with water. For instance, salts of weak acids or basis hydrolyse in aqueous solution, as in

$$Na^{+-}OOCCH_3 + H_2O \rightleftharpoons Na^+ + OH^- + CH_3COOH$$

The reverse reaction of esterification is another example. See *also* solvolysis.

Hydrophilic. Water attracting.

Hydrophobic. Lacking affinity for water. Water repelling.

Hydrosol. A sol in which the continuous phase is water.

Hydroxide. A metallic compound containing the ion OH^- (*hydroxide ion*) or containing the group—OH (hydroxyl group) bound to a metal atom. Hydroxides of typical metals are basic ; those of metalloids are amphoteric.

Hygroscopic. Describing a substance that absorbs moisture from the atmosphere.

Hyperfine Structure. See fine structure.

Hypertonic Solution. A solution that has a higher osmotic pressure than some other solution. *Compare* hypotonic solution.

Hypotonic Solution. A solution that has a lower osmotic pressure than some other solution. *Compare* hypertonic solution.

Hysteresis. Processes which occur mainly in solids in which the response obtained on increasing the variable differs from that on reducing the variable. Occurs *e.g.* during the magnetization of iron, during the absorption of moisture by textiles.

I

Ice. The solid form of water. The transition point between liquid and solid water at one atmosphere pressure is defined as 0° on the centrigrade scale, also equal to 0° celsius. This freezing temperature is lowered by increasing the pressure. By crystallization of water under greatly increased pressure it is possible to obtain at least nine forms of solid water which are physically distinct from ordinary ice (Ice I). These rank as distinct phases and their equilibrium relationships have been studied. The molecules in ice are held together by hydrogen bonds.

Ice Point. The temperature at which there is equilibrium between ice and water at standard atmospheric pressure (*i.e.*, the freezing or melting point under standard conditions). It was used as a fixed point (0°) on the celsius scale, but the kelvin and the International Practical Temperature Scale are based on the triple point of water.

Icosahedral. Having the symmetry of an icosahedron, a solid figure with twelve apices ; each face is triangular. Icosahedral units are found in many boron derivatives, *e.g.*, $B_{12}H_{12}^{2-}$. (See *diagram*).

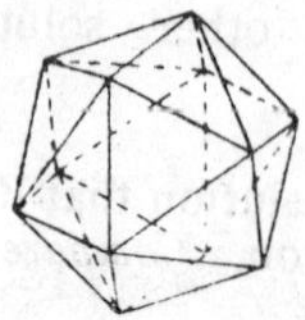

Ideal Gas (Perfect Gas). A hypothetical gas that obeys the gas laws exactly. An ideal gas would consist of molecules that occupy negligible space and have negligible forces between them. All collisions made between molecules and the walls of the container or between molecules and other molecules would be perfectly elastic, because the molecules would have no means of storing energy except as translational kinetic energy.

An ideal gas obeys exactly the gas laws :

$$PV = nRT$$

In practice, no gas shows ideal behaviour, although helium, hydrogen and nitrogen approximate to ideality at high temperatures and low pressures.

Ideal Solution. Refers to a solution, liquid or solid, in which the thermodynamic activity of each component is proportional to its mole fraction, *i e.*, a solution obeying Raoult's law. Such a solution would be formed from its components with zero heat of mixing, zero volume change and the ideal entropy of mixing. Only mixtures containing closely similar substances, *e.g.*, isotopic elements, form ideal solutions, although many mixtures show behaviour which approximates to ideality (*e.g.*, benzene and toluene).

Ignition Temperature. Refers to the temperature to which a substance must be heated before it will burn in air.

Imperial Units. The British system of units based on the pound and the yard. The former f.p.s. system was used in engineering and was loosely based on Imperial units ; for all scientific purposes SI units are now used. Imperial units are also being replaced for general purposes by metric units.

Implosion. An inward collapse of a vessel, especially as a result of evacuation.

Improper Rotation, S_n. A symmetry element consisting of rotation about an axis through $2\pi/n$ followed by reflection through a plane perpendicular to the axis of rotation.

Incandescence. The emission of light by a substance as a result of raising it to a high temperature.

Inelastic Neutron Scattering. Bombardment of a crystal by a beam of monochromatic neutrons results in a scattering of the latter due to collision with the crystal nuclei. Since the nuclei are not of infinite mass and they indulge in limited motion because of their thermal energy, such collisions cannot be regarded as truly elastic. The impinging neutron may either transfer some of its energy to the crystal vibrations or gain energy from these vibrations. Such collisions involving energy transfer are said to be inelastic and they result in scattered neutrons which are of different wavelength from the incident beam. The inelastic scattering of neutrons gives valuable information regarding crystal structure and vibrations within the crystal.

Infrared (IR). Electromagnetic radiation with longer wavelengths than visible radiation. The wavelength range is approximately 0.7 μm to 1 mm. Many materials transparent to visible light are opaque to infrared, including glass. Rock salt, quartz, germanium, or polyethylene prisms and lenses are suilable for use with infrared. Infrared radiation is produced by movement of charges on the molecular scale, *i.e.*, by vibrational or rotational motion of molecules. Of particular importance in chemistry is the absorption spectrum of compounds in the infrared region. Certain bonds between pairs of atoms (C$-$C, C$=$C, C$=$O, etc.) have characteristic vibrational frequencies, which correspond to bands in the infrared spectrum. Infrared spectra are thus used in finding the structures of new organic compounds. They are also used to 'fingerprint' and thus identify known compounds. At shorter wavelengths, infrared absorption corresponds to transitions between rotational energy levels, and can be used to find the dimensions of molecules (by their moment of inertia). See also electromagnetic radiation.

Inhibition. A reduction in the rate of a catalysed reaction by substances called *inhibitors*. In biochemical reactions, in which the catalysts are enzymes, if the inhibitor molecules resemble the substrate molecules they may bind to the active site of the enzyme, so preventing normal enzymatic activity. Alternatively they may form a complex with the substrate-enzyme intermediate or irreversibly destroy the enzyme configuration and active-site properties. The toxic effects of many substances are produced in this way. Inhibition by reaction products (*feed-back inhibition*) is important in the control of enzyme activity.

Insulator. Refers to a substance with a very low electrical conductivity. In insulators there is a wide separation between completely filled and completely empty electronic energy levels. Most pure solid ionic substances are insulators although impurities or imperfections may introduce semi-conductivity.

Internal Conversion. Refers to a process in which an excited atomic nucleus decays to the ground state and the energy released is transferred by electromagnetic coupling to one of the bound electrons of that atom rather than being released as a photon. The coupling is generally with an electron in the K-, L-, or M-shell of the atom, and this *conversion electron* is ejected from the atom with a kinetic energy equal to the difference between the nuclear transition energy and the binding energy of the electron. The resulting ion is itself in an excited state and usually subsequently emits an Auger election or an X-ray photon.

Internal Energy (E). For any system in a given state it is possible to ascribe to it a definite amount of energy. This energy is termed the internal energy of the system and includes all forms of energy other than those arising from the position of the system in space. The internal energy (E) is related to the enthalpy (H) by the equation

$$H=E+PV$$

where P is the pressure and V the volume.

Internal Resistance. Resistance of a source of electricity. In the case of a cell, when a current is supplied, the potential difference between the terminals is lower than the e.m.f. The difference (*i e.*, e.m.f.-p.d.) is proportional to the current supplied. The internal resistance (r) is given by :

$$r=(E-V)/I$$

where E is the e.m.f., V the potential difference between the terminals, and I the current.

Interstitial. See defect.

Inhibitor. A substance that lowers the efficiency of a catalyst, *i.e.*, slows down the rate of reaction. Hydrogen sulphide, hydrogen cyanide, metcury salts, and arsenic compounds readily inhibit heterogeneous catalysts by adsorption. For example, arsenic compounds inhibit platinum catalysts in the oxidation of sulphur (IV) oxide to sulphur (VI) oxide. Inhibitors must not be confused with negative catalysts ; inhibitors do not change the pathway of a reaction.

Invariant System. A system which, when at equilibrium, possesses no degrees of freedom. See phase rule.

Inversion Temperature. Most gases when rapidly expanded experience a cooling effect (the Joule-Thomson effect). Below 193K hydrogen also shows this effect, but above this temperature it behaves abnormally and is warmed by expansion. This temperature of 193K is called the inversion temperature. Helium shows similar behaviour to hydrogen with an inversion temperature of 33K.

Ion. Refers to an atom or group of atoms that has either lost one or more electrons, making it positively charged (an anion), or gained one or more electrons, making it negatively charged (a cation).

Ionic Atmosphere. The Debye-Huckel theory of interionic attraction shows that in solutions ions of one charge type are surrounded by an atmosphere of ions predominately of the opposite charge type. The effect felt by the central ion is the same in all directions. When an electric field is applied in the solution, however, the central ion moves in one direction whereas the ionic atmosphere moves in the opposite sense giving rise to an asymmetric distribution of ions about the central ion. This effect is important in studies of electrical conductivities of solutions.

Ionic Crystal. See crystal.

Ionic Product. The product of the concentrations of the ions of a substance in solution. This factor is a constant for weakly ionized substances, because the concentration of undissociated material may be taken to be constant, if it is present in a large excess.

For a sodium chloride solution the ionic product is $[Na^+][Cl^-]$; for a calcium chloride solution it is $[Ca^{2+}][Cl^-]^2$. In pure water, there is an equilibrium with a small amount of self-ionization :

$$H_2O \rightleftharpoons H^+ + OH^-$$

The equilibrium constant of this dissociation is given by

$$K_w = [H^+][OH^-]$$

since the concentration $[H_2O]$ can be taken as constant. K_w is referred to as the ionic product of water. It has the value 10^{-14} mol^{2}dm^{-6} at 25°C. In pure water (*i.e.*, no added acid or added alkali) $[H^+]=[OH^-]=10^{-7}$ mol dm^{-3}.

Ionic Strength. For an electrolyte solution the ionic strength is given by

$$I = \tfrac{1}{2}\Sigma_i m_i z_i^2$$

where z is the charge on the ionic species and m is the modal concentration, the summation being continued over all the different ionic species in the solution.

Ionium, ^{230}Th. A former specific name for a Th isotope.

Ionization. Refers to the process of producing ions. There are several ways in which ions may be formed from atoms or molecules. In certain chemical reactions ionization occurs by transfer of electrons ; for example, sodium atoms and chlorine atoms react to form sodium chloride, which consists of sodium ions (Na^+) and chloride ions (Cl^-). Certain molecules can ionize in solution ; acids, for example, form hydrogen ions as in the reaction

$$H_2SO_4 \rightarrow 2H^+ + SO_4^{2-}$$

The 'driving force' for ionization in a solution has been solvation of the ions by molecules of the solvent. H^+, for example, is solvated as a hydronium ion, H_3O^+.

Ions can also be produced by ionizing radiation ; *i.e.*, by the impact of particles or photons with sufficient energy to break up molecules or detach electrons from atoms : $A \rightarrow A^+ + e^-$. Negative ions can be formed by capture of electrons by atoms or molecules : $A + e^- \rightarrow A^-$.

Ionization Chamber. A device for measuring the absolute intensity of a beam of X-rays or ionizing particles. The rays or particles are allowed to ionize the vapour contained in the chamber, which also contains suitably placed and charged electrodes. The intensity of the beam is measured by measuring the current flowing between the electrodes.

Ionization Gauge. A vacuum gauge consisting of a three-electrode system inserted into the container in which the pressure is to be measured. Electrons from the cathode are attracted to the grid, which is positively biased. Some pass through the grid but do not reach the anode, as it is maintained at a negative potential. Some of these electrons do, however, collide with gas molecules, ionizing them and converting them to positive ions. These ions are attracted to the anode, the resulting anode current can be used as a measure of the number of gas molecules present. Pressure as low as 10^{-6} pascal can be measured in this way.

Ionization, Heat of. The amount of heat required to split one mole of an electrolyte into its constituent ions. For water, the heat of ionization is 57.3 kJ mol^{-1}.

Ionization of Water. Pure water has a very low conductivity and is only very slightly ionized. The concentrations of hydrogen and hydroxyl ions, which are equal, are 10^{-7}g, ions l^{-1}. The ionic product for water is $[H^+][OH^-] = 10^{-14}$ at 25°C, but increases rapidly with temperature. The pH of pure water is clearly 7. Any solution with pH$>$7 contains excess hydroxyl ions and is alkaline ; those with pH$<$7 are acidic, containing excess hydrogen ions.

Ionizing Radiation. Radiation of sufficiently high energy to cause ionization in the medium through which it passes. It may consist of a stream of high-energy particles (*e.g.*, electrons, protons, alpha-particles) or shortwavelength electromagnetic radiation (ultraviolet, X-rays, gamma-rays). This type of radiation can cause extensive damage to the molecular structure of a substance either as a result of the direct transfer of energy to its atoms or molecules or as a result of the secondary electrons released by ionization. In biological tissue the effect of ionizing radiation can be very serious, usually as a consequence of the ejection of an electron from a water molecule and the oxidizing or reducing effects of the resulting highly reactive species :

$$H_2O \rightarrow e^- + H_2O + H_2O^+ \rightarrow OH + H_3O^+ + H,$$

where the dot before a radical indicates an unpaired electron and an denotes an excited species.

Ionogenic Surface. The stability of a colloid particle is due to the charge it carries. This charge is ionic in nature and if the stabilizing ions on the surface are due to the ionization of the particle wall material, the surface is said to be ionogenic.

Ion-pair. The Debye-Huckel theory of interionic attraction in solution satisfactorily interprets the behaviour of very dilute electrolyte solutions where long-range interionic effects are important. When the ions are close together, the approximations made in the theory no longer hold and the energy of the mutual attraction of ions is great enough to give rise to the formation of ion-pairs in solution which are capable of withstanding collisions with solvent molecules. This effect is more pronounced in mixed and non-aqueous solvent systems and there is evidence for triple ion-pairing in some instances.

Ion-selective Electrodes. Refers to an electrode which responds specifically to one ion present in solution and develops a potential according to the concentration of that ion. Used for metals, *e.g.*, Zn^{2+}, Cu^{2+}, Hg^{2+}, Na^+, Li^+ and Ba^{2+}, etc. Also used for anions. The electrode contains a standard solution and a membrane or other conductor placed in contact with the solution to be examined. Glass electrodes are widely used ion-selective electrodes.

IR. Infrared.

Ion-microprobe Analysis. A technique which is used for analysing the surface composition of solids. The sample is bombarded with a narrow beam (as small as 2 μm diameter) of high-energy ions. Ions ejected from the surface by sputtering are detected by mass spectrometry. The technique allows quantitative analysis of both chemical and isotopic compositions for concentrations as low as a few parts per million.

Ion Pump. Refers to a type of vacuum pump that can reduce the pressure in a container to about 1 nanopascal by passing a beam of electrons through the residual gas. The gas gets ionized and the positive ions formed are attracted to a cathode within the container where they remain trapped. The pump is only useful at very low pressures, *i.e.*, below about 1 micropascal. The pump is having a limited capacity because the absorbed ions eventually saturate the surface of the cathode. A more effective pump can be made by simultaneously producing a film of metal by ion impact (sputtering), so that fresh surface is continuously produced. The device is then known as a *sputter-ion pump*.

Irreversible Reaction. A reaction in which conversion to products is complete ; *i.e.*, there is little or no back reaction.

Isentropic Process. Refers to any process that takes place without a change of entropy. The quantity of heat transferred, δQ, in a reversible process is proportional to the change in entropy, δS, *i.e.*, $\delta Q = T\delta S$, where T is the thermodynamic temperature. Therefore, a reversible adiabatic process is isentropic, *i.e.*, when $\delta Q = 0$, δS also equals 0.

Isobar. A curve on a graph indicating readings taken at constant pressure.

Isodispersion. Many natural substances of high molecular weight form sols in which the dispersed particles, of colloidal dimensions, are all of the same size, and thus are termed isodisperse. Thus haemoglobin forms an isodispersion of particles all of mol. wt. 68,000 Helix haemocyanin forms an isodispersion of mol. wt. 5,000,000.

Artificially prepared colloidal sols usually have articles of widely varying size, but approximation to isodispersion can be produced by differential ultra-filtration.

Isomorphism. The existence of two or more substances (*isomorphs*) that have the same crystal structure, so that they are able to form solid solutions.

Isopiestic. Solutions of salts, etc. which have the same partial pressure of solvent are termed isopiestic solutions. The isopiestic method of studying the osmotic properties of solutions is to allow solutions to come to isopiestic equilibrium, by diffusion of vapour from one to another, and with a solution of a standard substance in an enclosed vessel at a constant temperature ; the concentrations of the isopiestic solutions are then determined. The method is slow but capable of attaining a high accuracy.

Isotherm. Refers to a line on a chart or graph joining points of equal temperature.

Isothermal Change. A process that takes place at a constant temperature. Throughout an isothermal process, the system is in thermal equilibrium with its surroundings. For example, a cylinder of gas in contact with a constant-emperature box may be compressed slowly by a piston. The work done appears as energy, which flows into the reservoir to keep the gas at the same temperature. Isothermal changes are contrasted with *adiabatic changes*, in which no energy enters or leaves the system, and the temperature changes. In practice no process is perfectly isothermal and none is perfectly adiabatic, although some can approximate in behaviour to one of these ideals.

Isothermal Process. Any process that takes place at constant temperature. In such a process heat is, if necessary, supplied or removed from the system at just the right rate to maintain constant temperature.

Isotones. Two or more nuclides that have the same neutron numbers but different proton numbers.

Isotonic. Describing solutions that have the same osmotic pressure.

Isotopes. Two or more species of the same element differing in their mass numbers because of differing numbers of neutrons in their nuclei. The nuclei must have the same number of protons (an element is characterized by its proton number). Isotopes of the same element have very similar properties because they have the same electron configuration, but differ slightly in their physical properties. An unstable isotope is termed a *radioactive isotope* or *radioisotope*.

Isotope Effect. Reactions involving the breaking of a C—H bond are slower when that hydrogen is replaced by deuterium or tritium and are distinguished by a 'primary kinetic isotope effect'. This difference in rate of breaking of a C—H bond compared to a C—D bond is due to the greater vibrational energy at room temperature of a C—H bond by comparison with that of the C—D bond. Consequently, less energy is required to break the C—H bond.

Isotopic Number. Refers to the difference between the number of neutrons in an atom and the number of protons.

Isotropic. A substance which possesses identical properties in all directions in its crystal. Cubic substances and some non-crystalline glasses are isotropic.

J

Jj Coupling. See Russell-Saunders coupling.

Joule. Symbol : J. The SI unit of energy and work, equal to the work done when the point of application of a force of one newton moves one metre in the direction of action of the force. 1 J=1 Nm. The joule is the unit of all forms of energy.

Joule's Law. The internal energy of a given mass of gas is independent of its volume and pressure, being a function of temperature alone. This law applies only to ideal gases (for which it provides a definition of thermodynamic temperature) as in a real gas intermolecular forces would cause changes in the internal energy should a change of volume occur. See also Joule-Thomson effect.

Joule-Thomson Effect, Joule-Kelvin Effect. Most gases (except H and He) undergo cooling when they are expanded adiabatically through a throttle. This is called the Joule-Thomson effect. The change in temperature is proportional to the drop in pressure ; the Joule-Thomson coefficient is the change of temperature per unit change of pressure. The effect, which is used for liquefying air, making solid carbon dioxide, etc. is related to departures from Boyle's law and Joule's law. H and He, which show a warming effect at ordinary temperatures, also show cooling if the experiment is conducted below thier inversion temperatures.

K

Kapustinskii Equation. For an ionic crystal composed of cations and anions, of respective charge z_+ and z_-, which behave as hard spheres, the lattice energy (U) may be obtained from the expression

$$U = \frac{Lz_+z_-e^2M}{R}\left(1-\frac{1}{n}\right);$$

where L is Avagadro's number ; e, the electronic charge ; R, the equilibrium internuclear distance ; M, the Madelung constant and n an integer. The Kapustinskii equation is a simplification of the above expression in which it is assumed that the value of M/v, where v is the number of ions in one 'molecule', is constant for different structures and that n has an average value of 9. Thus, if r_+ and r_- are the thermodynamic radii of cation and anion respectively, and $R=(r_++r_-)$, the Kapustinskii equation states

$$U = \frac{0.777N_0vz_+z_-e^2}{(r_++r_-)}.$$

Katharometer. An instrument for comparing the thermal conductivities of two gases by comparing the rate of loss of heat from two heating coils surrounded by the gases. The instrument can be used to detect the presence of a small amount of an impurity in air and is also used as a detector in gas chromatography.

Kayser. A suggested unit of frequency 1 cm^{-1}. The kilo-kayser is 1000 cm^{-1}.

Kelvin. Symbol : K. The SI base unit of thermodynamic temperature. It is defined as the fraction 1/273.16 of the thermodynamic temperature of the triple point of water. Zero kelvin (0 K) is absolute zero. One kelvin is the same as one degree on the Celsius scale of temperature.

Kelvin Effect. See Thomson effect.

Kelvin Scale. See absolute temperature.

Kilo. Symbol : k. A prefix denoting 10^3. For example, 1 kilometre (km)$=10^3$ metres (m).

Kilogram. Symbol kg. The SI unit of mass defined as a mass equal to that of the international platinum-iridium prototype kept by the International Bureau of Weights and Measures at Sevres, near Paris.

Kilogramme. An alternative spelling of ***kilogram***.

Kilowatt-Hour. Symbol : kWh. A unit of energy, usually electrical, equal to the energy transferred by one kilowatt of power in one hour. It has a value of 3.6×10^6 joules.

Kinematic Viscosity. A common method of determining viscosity is to time the flow of a liquid through a capillary under its own gravity. This gives readings in terms of kinematic viscosity (Stokes) which is equivalent to dynamic viscosity (Poises) divided by density. The SI unit is $m^2\ s^{-1}$.

Kinetics, Chemical. The branch of chemistry which is concerned with the explanation of observed characteristics of chemical reactions (*e.g.*, the variation of reaction velocities with pressure, temperature or concentration).

Kinetic Effect. A chemical effect that depends on reaction rate rather than on thermodynamics. For example, diamond is thermodynamically less stable than graphite ; its apparent stability depends on the vanishingly slow rate at which it is converted. Overvoltage in electrolytic cells is another example of a kinetic effect. *Kinetic isotope effects* are changes in reaction rates produced by isotope substitution. For example, if the slow step in a chemical reaction is the breaking of a C—H bond, the rate for the deuterated compound would be slightly lower because of the lower vibrational frequency of the C—D bond. Such effects are used in investigating the mechanisms of chemical reactions.

Kinetic Energy. See energy.

Kinetic Isotope Effect. See isotopes.

Kinetic Theory. A theory, largely the woɩk of Count Rumford (1753-1814), James Joule (1818-89), and James Clerk Maxwell (1831-79), that explains the physical properties of matter in terms of the motions of its constituent particles. In a gas, for example, the pressure is due to the incessant impacts of the gas molecules on the walls of the container. If it is assumed that the molecules occupy negligible space, exert negligible forces on each other except during collisions, are perfectly elastic, and make only brief collisions with each other, it can be shown that the pressure p exerted by one mole of gas containing n molecules each of mass m in a container of volume V, will be given by :

$$p=nmc^2/3V,$$

where c^2 is the mean square speed of the molecules. As according to the gas laws for one mole of gas : $pV=RT$, where T is the thermodynamic temperature, and R is the molar gas constant, it follows that :

$$RT=nmc^2/3$$

Thus, the thermodynamic temperature of a gas is proportional to the mean square speed of its molecules. As the average kinetic energy of translation of the molecules is $mc^2/2$, the temperature is given by :

$$T=(mc^2/2)(2n/3R)$$

The number of molecules in one mole of any gas is the Avogadro constant, N_A ; therefore in this equation $n=N_A$. The ratio R/N_A is a constant called the Boltzmann constant (k.)

The average kinetic energy of translation of the molecules of one mole of any gas is therefore $3kT/2$. For monatomic gases this is proportional to the internal energy (U) of the gas, *i.e.*,

$$U = N_A 3kT/2$$

and as $k = R/N_A$

$$U = 3RT/2$$

For diatomic and polyatomic gases the rotational and vibrational energies also have to be taken into account (see degrees of freedom).

In liquids, according to the kinetic theory, the atoms and molecules still move around at random, the temperature being proportional to their average kinetic energy. However, they are sufficiently close to each other for the attractive forces between molecules to be important. A molecule that approaches the surface will experience a resultant force tending to keep it within the liquid. It is, therefore, only some of the fastest moving molecules that escape ; as a result the average kinetic energy of those that fail to escape is reduced. In this way, evaporation from the surface of a liquid causes its temperature to fall.

In a crystalline solid the atoms, ions, and molecules are able only to vibrate about the fixed positions of a crystal lattice ; the attractive forces are so strong at this range that no free movement is possible.

Kirchhoff's Equation. If Q is the heat evolved when a process (physical or chemical) is carried out at temperature T, the heat which would be evolved if the same process were carried out at a different temperature can be calculated with the aid of Kirchhoff's equation, namely

$$dQ/dT = C_1 - C_2$$

where C_1 and C_2 are the total heat capacities of the system before and after the process respectively (*e.g.*, reactants and products in a chemical reaction). Kirchhoff's equation is a direct consequence of the first law of thermodynamics.

Kohlrausch Equation. This equation, which describes the behaviour of strong electrolytes on dilution, states that

$$\Lambda_\infty - \Lambda_v = kC^{\frac{1}{2}}$$

where Λ_∞ is the equivalent conductivity at infinite dilution, Λ_v that at volume v, k is a constant and C is the concentration of electrolyte. The equation is valid only at high dilutions.

Kohlrausch's Law. If a salt is dissolved in water, the conductivity of the (dilute) solution is the sum of two values—one depending on the positive ions and the other on the negative ions. The law, which depends on the independent migration of ions, was deduced experimentally by the German chemist Friedrich Kohlrausch (1840-1910).

L

Lambda Point, λ Point. The temperature of transition between He_I and HE_{II}.

Lambert's Law. This states that layers of equal thickness of a homogeneous marerial absorb equal proportions of light

$$I = I_0 e^{-K^d},$$

where I is the intensity of the transmitted light, I_0 that of the incident light, d the thickness of the layer and K a constant, charactsristic of the substance, known as the absorption coefficient ; K also depends on the wavelength of the light employed. When solutions are considered, it is clearly desirable to modify this expression to include the concentration of the absorbing molecules. This modification is embodied in Beer's law.

Lamb Shift, A small energy difference between two levels ($^2S_{1/2}$ and $^2P_{1/2}$) in the hydrogen spectrum. The shift results from the quantum interaction between the atomic electron and the electromagnetic radiation. It was first explained by Willis Eugene Lamb (1913-).

Lamp Black. A finely divided (microcrystalline) form of carbon made by burning organic compounds in insufficient oxygen.

Lande g Factor. See g factor.

Langmuir. An arbitrary unit used as a measure of adsorption. 10^{-6} torr adsorbant exposed to a surface for 1 second.

Langmuir Adsorption Isotherm. A theoretical equation, derived from the kinetic theory of gases, which relates the amount of gas adsorbed at a plane solid surface to the pressure of gas in equilibrium with the surface. In the derivation it is assumed that the adsorption is restricted to a mono-layer at the surface, which is considered to be energetically uniform. It is also assumed that there is no interaction between the adsorbed species. The equation shows that at a gas pressure, p, the fraction, θ, of the surface covered by the adsorbate is given by :

$$\theta=\frac{bp}{1+bp}$$

where b is a constant called the adsorption coefficient, it is the equilibrium constant for the adsorption process.

Although still used the Langmuir equation is only of limited value since in practice surfaces are energetic in homogeneous and interactions between adsorbed species often occur.

Latent Heat. The heat absorbed or evolved during a change in the physical state of a substance. The heat effect in such changes is reversible. The heat absorbed when a solid is converted to liquid at the m.p. is called the latent heat of fusion, that absorbed when a liquid is converted to a gas at the b.p. and 1 atmosphere pressure is termed the latent heat of vaporization. Latent heats are usually measured in joules per mole of substances. Latent heats are entropies of phase changes.

Latent Image. The change occurring in silver halide grains which makes them developable after exposure to light.

Lattice. A regular three-dimensional arrangement of points. A lattice is used to describe the positions of the particles (atoms, ions, or molecules) in a crystalline solid. The lattice structure can be examined, by X-ray diffraction techniques.

Lattice Energy. A measure of the stability of a crystal lattice, given by the energy that would be released per mole if atoms, ions, or molecules of the crystal were brought together from infinite distances apart to form the lattice. See Born-Haber cycle.

Lattice Vibrations. The periodic vibrations of the atoms, ions, or molecules in a crystal lattice about their mean positions. On heating, the amplitude of the vibrations increases until they are so energetic that the lattice breaks down. The temperature at which this happens is the melting point of the solid and the substance becomes a liquid. On cooling, the amplitude of the vibrations diminishes. At absolute zero a residual vibration persists, associated with the zero-point energy of the substance. The increase in the electrical resistance of a conductor is due to increased scattering of the free conduction electrons by the vibrating lattice particles.

Laue Pattern. The symmetrical array of spots obtained on a photographic plate exposed to a non-homogeneous beam of X-rays after its passage through a crystal. The patterns constitute the earliest, although one of the most difficult, methods of investigating crystal structure by means of X-rays.

Laves Phases. Alloys, *e.g.*, $MgZn_2$, where structure is determined by size rather than by electronic configurations.

Law of Isomorphism. See Mitscherlich's law.

Law of Mass Action. At constant temperature, the rate of a chemical reaction is directly proportional to the active mass of the reactants, the active mass being taken as the concentration in mol dm^{-3}. In general, the rate of reaction decreases steadily as the reaction proceeds; a measure of the concentration of any of the reactants will give a measure of the rate of reaction.

For the reaction $A+B \rightarrow$ products, the law of mass action states that

$$\text{rate}=k[A][B]$$

where [A] represents the concentration of A in mol dm^{-3}. The term *active mass* can only equal the concentration in mol dm^{-3} if there is no interaction or interference between the reacting molecules. Many systems exhibit interactions and interference and consequently the concentration in mol dm^{-3} has to be multiplied by an activity coefficient in order to obtain the effective active mass.

Layer Lattice. Crystal structures in which well-defined layers, either simple or composite, of atoms may be distinguished. The forces between the atoms within the layers are stronger than those holding the layers together, leading to good cleavage parallel to the layers. *e.g.*, graphite, boric acid, cadmium iodide and chloride, and molybdenum sulphide.

Lead-Acid Accumulator. A type of electrical accumulator used in vehicle batteries. It has two sets of plates : spongy lead plates connected in series to the negative terminal and lead oxide plates connected to the positive terminal. The material of the electrodes is held in a hard lead-alloy grid. The plates are interleafed. The electrolyte is dilute sulphuric acid.

The e.m.f. when fully charged is about 2.2 V. This falls to a steady 2 V when current is drawn. As the accumulator begins to run down, the e.m.f. falls further. During discharge the electrolyte becomes more dilute and its relative density falls. To recharge the accumulator, charge is passed through it in the opposite direction to the direction of current supply. This reverses the cell reactions and increases the relative density of the electrolyte (*c* 1.25 for a fully charged accumulator).

The electrolyte contains hydrogen ions (H^+) and sulphate ions (SO_4^{2-}). During discharge, H^+ ions react with the lead (IV) oxide to give lead(II) oxide and water

$$PbO_2+2H^++2e^- \rightarrow PbO+H_2O$$

This reaction takes electrons from the plate, causing the positive charge. There is a further reaction to yield soft lead sulphate :

$$PbO+SO_4^{2-}+2H^+ \rightarrow PbSO_4+H_2O+2e^-$$

Electrons are released to the electrode, producing the negative charge.

During charging the reactions are reversed :

$$PbSO_4+2e^- \rightarrow Pb+SO_4^{2-}$$
$$PbSO_4+2H_2O \rightarrow PbO_2+4H^++SO_4^{2-}+2e^-$$

Le Chatelier's Theorem or Principle. When a constraint is applied to a system in equilibrium, the equilibrium will tend to move in such a way as to neutralize the effect of the constraint. Thus, *e g.*, in a reaction such as

$$N_2O_4 \rightleftharpoons 2NO_2$$

increasing the total pressure will cause the equilibrium to move towards the formation of more N_2O_4 thereby decreasing the total number of molecules in the system.

Leclanche Cell. A primary voltaic cell consisting of a carbon rod (the anode) and a zinc rod (the cathode) dipping into an electrolyte of a 10-20% solution of ammonium chloride. Polarization is prevented by using a mixture of manganese dioxide mixed with crushed carbon, held in contact with the anode by means of a porous bag or pot ; this reacts with the hydrogen produced. This wet form of the cell, devised in 1867 by Georges Leclanche (1839-82), has an e.m.f. of about 1.5 volts. The dry cell based on it is widely used in torches, radios, and calculators.

Light (Visible Radiation). A form of electromagnetic radiation able to be detected by the human eye. Its wavelength range is between about 400 nm (far 'red') and about 700 nm (far 'violet'). The boundaries are not precise as individuals vary in their ability to detect extreme wavelengths ; this ability also declines with age.

Light-Heavy Selectivity. In solvent extraction the phenomenon in which the extractive power of the solvent is dependent on the molecular weight of the component extracted. *E.g.*, in the removal of aromatic compounds from wide molecular weight-range petroleum fractions some of the non-aromatic material may dissolve before the higher aromatics. Precipitation metods, making use of the light-heavy selectivity, are often employed. See deasphalting.

Limiting Density (of a Gas). If a mass w g of a gas occupies v litres at 0° and pressure p atmospheres, the quotient w/pv is the density of the gas per unit pressure. Since gases do not obey Boyle's law accurately, w/pv varies with the pressure. The value of this quotient when p approaches zero is called the limiting density of the gas. It represents the density which the gas would possess as an ideal gas, *i.e.*, a gas which obeys the gas laws perfectly.

Line Spectra. Absorption of energy can only occur between allowed energy levels. Absorption of energy by an atom can only lead to specific changes to excited states (*e.g.*, with transfer of electrons from one energy level to others). The reverse of this process is generally accompanied by emission of energy of a single definite frequency corresponding to the energy involved in the process. The spectrum of an atom thus consists of a series of discrete lines, each line at a frequency corresponding to the appropriate electronic transition. The line spectrum is characteristic of an element in its atomic state.

Liquefaction of Gases. The conversion of a gaseous substance into a liquid. This is usually achieved by one of four methods or by a combination of two of them :

(1) by vapour compression, provided that the substance is below its critical temperature ;

(2) by refrigeration at constant pressure, typically by cooling it with a colder fluid in a countercurrent heat exchanger ;

(3) by making it perform work adiabatically against the atmosphere in a reversible cycle :

(4) by the Joule-Thomson effect.

Large quantities of liquified gases are now used commercially, especially liquefied petroleum gas and liquefied natural gas.

Liquid. A phase of matter between that of a crystalline solid and a gas. In a liquid, the large-scale three-dimensional atomic (or ionic or molecular) regularity of the solid is absent but, on the other hand, so is the total disorganization of the gas. Although liquids have been studied for many years there is still no comprehensive theory of the liquid state. It is clear, however, from diffraction studies that there is a short-range structural regularity extending over several molecular diameters. These bundles of ordered atoms, molecules, or ions move about in relation to each other, enabling liquids to have almost fixed volumes, which adopt the shape of their containers.

Liquid Crystals. On heating a number of substances, *e.g.*, *p*-azoxyanisole and ammonium oleate, a cloudy liquid is first formed which changes at higher temperature to a clear liquid, each transition occurring at a fixed temperature. The cloudy liquid has definite ordered structure and it is called a liquid crystal, or more generally, is said to be in a mesamorphic state; its viscosity, etc., are those of liquids. Substances which form liquid crystals are composed of molecules which possess a high degree of asymmetry—long thin molecules or flat planar ones—which tends to allow little rotation in the liquid state at low temperature. Liquid crystals exhibiting two-dimensional order are said to be in the cholesteric or smetic phases; those showing only one-dimensional order are nematic. Application of a voltage causes disorder; the basis of the extensive use of liquid crystals in display units. Cholesteric liquids scatter coloured, polarized light and can be made the basis of thermometer units. Spectroscopic studies in the presence of liquid crystals give much additional information over more simple techniques.

Liquids, Structure of. It was originally thought that liquids and gases were closely similar, the molecules being arranged in a random manner. However, X-ray and electron diffraction and neutron scattering studies show that the structure of a liquid is intermediate between the perfect disorder of molecules in a gas and the highly perfect three-dimensional order in a solid. The immediate environment of an atom or molecule in a liquid is not very different from that in a solid and the density of packing is only slightly lower, but the order is very local and no long-range order occurs.

Liquidus Curve. The freezing point of a molten mixture of substances varies with the composition of the mixture. If the freezing points are plotted as a function of the composition, the line joining the points is called a liquidus curve. Such mixtures usually freeze over a range of temperature. If the temperature at which the last traces of liquid just solidify (assuming that sufficient time has been allowed for equilibrium to be established) are plotted against composition the resulting line is called a solidus curve.

Litre. Symbol : *l*. A unit of volume now defined as 10^{-3} metre^3; *i.e.*, 1000 cm^3. The millilitre (ml) is thus the same as the cubic centimetre (cm^3). However, the name is not recommended for precise measurements since the litre was formerly defined as the volume of one kilogram of pure water at 4°C and standard pressure. On this definition, one litre is the same as 1000.028 cubic centimetres.

Logarithmic Scale. 1. A scale of measurement in which an increase or decrease of one unit represents a tenfold increase or decrease in the quantity measured. Decibels and pH measurements are common examples of logarithmic scales of measurement. 2. A scale on the axis of a graph in which an increase of one unit represents a tenfold increase in the variable quantity. If a curve $y=x^n$ is plotted on graph paper with logarithmic scales on both axes, the result is a straight line of slope n, *i.e.*, $\log y = n \log x$, which enables n to be determined.

Loschmidt's Constant (Loschmidt Number. The number of particles per unit volume of an ideal gas at STP. It has the value 2.68719×10^{25} m^{-3} and was first worked out by Joseph Loschmidt (1821-95).

Low Energy Electron Diffraction (LEED). A technique used to investigate the two-dimensional surface structure of solids. The surface under investigation is bombarded with electrons of low energy, typically between 6 and 600 V, and the electrons diffracted by the surface atoms are collected on a fluorescent screen. Using this method it is possible to examine both surface structure and rearrangements which may occur following chemi-sorption and surface reaction.

Lowering of Vapour Pressure. A colligative property of solutions in which the vapour pressure of a solvent is introduced. When both solvent and solute are volatile the effect of increasing the solute concentration is to lower the partial vapour pressure of each component. When the solute is a solid of negligible vapour pressure the lowering of the vapour pressure of the solution is directly proportional to the number of species introduced rather than to their nature and the proportionality constant is regarded as a general solvent property. Thus the introduction of the same number of moles of any solute causes the same lowering of vapour pressure, if dissociation does not occur. If the solute dissociates into two species on dissolution the effect is doubled.

The kinetic model for the lowering of vapour pressure treats the solute molecules as occupying part of the surface of the liquid phase and thereby restricting the escape of solvent molecules. The effect can be used in the measurement of relative molecular masses, particularly for large molecules, such as polymers. See *also* Raoult's law.

Low-Spin State. See high spin state. The minimum number of unpaired electrons.

Low Temperature Baths. These are used for freezing and distilling gases. Liquid N_2 (−196°C), liquid O_2 (−183°C−very hazardous with any organic material), solid CO_2 (−78.5°C) are used together with various slush baths (liquid plus liquid N_2 or CO_2) −160°C isopentane ; −140°C, 30-60 petroleum ether ; −130°C pentane ; −126°C methylcyclohexane; −116°C diethyl ether ; −111°C CCl_3F ; −95°C toluene ; −60°C $CHCl{=}CCl_2$; −45°C C_6H_5Cl ; −23°C CCl_4.

Lumen. Symbol : lm. The SI unit of luminous flux equal to the flux emitted by a uniform point source of 1 candela in a solid angle of 1 steradian.

Luminescence. The emission of radiation from a substance in which the particles have absorbed energy and gone into excited states. They then return to lower energy states with the emission of electromagnetic radiation. If the luminescence persists after the source of excitation is removed it is called *phosphorescence* ; if not, it is called *fluorescence.*

Lux. Symbol : lx. The SI unit of illuminance equal to the illumination produced by a luminous flux of 1 lumen distributed uniformly over an area of 1 square metre.

Lyophilic. Having an affinity for a solvent ('solvent-loving ; if the solvent is water the term *hydrophilic* is used). See colloid.

Lyophobic. Lacking any affinity for a solvent ('solvent-hating' ; if the solvent is water the term *hydrophobic* is used). See colloid.

Lyophilic. Colloidal substances in which the dispersed particles display a marked stability due to solvation are said to be lyophilic. See colloids.

Lyophobic. Colloidal substances in which the dispersed particles are unsolvated and hence easily coagulated. See colloids.

Lyotropic Series. Although hydrophilic sols are not affected by small concentrations of electrolytes, they may be 'salted out' by the addition of certain salts which possess strong dehydrating properties. Citrates, tartrates and sulphates are very efficient in this connection ; iodides and thiocyanates tend to disperse rather than coagulate. Thus silk or cellulose will peptize to colloidal sols in strong thiocyanate solutions but are reprecipitated by the addition of sulphates to the sols. The arrangement of the different anions in order of their salting out efficiency is termed the lyotropic or Hofmeister series. This also has a connection with the coagulating power of salts on hydrophobic sols. See colloids, hydrophilic, hydrophobic, salting out.

M

Macleod's Equation

$$\gamma = K(D-d)^4$$

where γ is the surface tension, D the density of the liquid and d the density of the vapour.

Macromolecular Crystal. A crystal composed of atoms joined together by covalent bonds, which form giant three-dimensional or two-dimensional networks. Diamond is an example of a macromolecular crystal.

Macromolecule. Molecules of high molecular weight (usually greater than 10,000). Examples of this type of molecule are proteins and some polymers. The term macromolecular is also used to describe structures such as that of diamond, which may be regarded as being one large molecule.

Madelung Constant. For an ionic crystal composed of cations and anions of respective change z_+ and z_-, the lattice energy U_0 may be derived as the balance between the coulombic attractive and repulsive forces. This approach yields the Born-Lande equation,

$$U_0 = \frac{LMz_+z_-e^2}{r_0}\left(1-\frac{1}{n}\right),$$

where L is Avagadro's number ; r_0, the equilibrium internuclear distance ; *e*, the electronic charge ; *n*, an integer and M the Madelung constant.

Values of M depend upon the structure varying between 1.763 (CsCl) and 2.519 (fluorite structure).

Magnetic Moment. For paramagnetic substances the magnetic moment (μ) is related to the magnetic susceptibility (χ) by the equation

$$\mu=2.84\sqrt{\chi_m(T-\theta)}$$

where χ_m is the molar susceptibility, T the absolute temperature and θ is a constant for the particular substance (called the Curie temperature).

Magnetic Polarization of Light. When a beam of plane polarized light is passed through a transparent medium which is placed in a magnetic field, the plane of polarization is rotated. This phenomenon is known as magnetic rotation, magnetic optical rotatory cuspersion, the Faraday effect. The light beam must be travelling in the same direction as the magnetic lines of force. The angle of rotation (ω), is related to the length of the medium traversed (l) and to the magnetic field strength (H) by the equation : $w=\phi Hl$, where ϕ is a constant known as Verdet's constant. Magnetic rotation is not connected with normal optical activity. The effect is shown by transparent substances whether or not they possess a centre of asymmetry.

Magnetic Resonance. See nuclear magnetic resonance, electron spin resonance.

Magnetic Susceptibility. Atoms or molecules may have permanent magnetic moment or an induced moment due to the influence of a magnetic field. In general, the extent of magnetization of any substance is a function of the field in which that substance is placed. The magnetic susceptibility per unit volume, *k*, is defined as the intensity of magnetization induced (I) divided by the field strength (H)

$$k=I/H.$$

The molar magnetic susceptibility χ_m is given by the expression

$$\chi_m = \frac{M}{\rho} k.$$

where M is the molecular weight and ρ the density of the substance. Magnetic susceptibilities are usually determined using a Gouy balance but nuclear magnetic resonance methods may also be used. For a paramagnetic substance $1/\chi$ is proportional to the absolute temperature (Curie-Weiss law).

Magnetochemistry. The branch of physical chemistry concerned with measuring and investigating the magnetic properties of compounds. It is used particularly for studying transition-metal complexes, many of which are paramagnetic because they have unpaired electrons. Measurement of the magnetic susceptibility allows the magnetic moment of the metal atom to be calculated, and this gives information about the bounding in the complex.

Magneton. A unit for measuring magnetic moments of nuclear, atomic, or molecular magnets. The *Bohr magneton* μ_b has the value of the classical magnetic moment of an electron, given by

$$\mu_b = eh/4\pi m_e = 9.274 \times 10^{-24} \text{ Am}^2,$$

where e and m_e are the charge and mass of the electron and h is the Planck constant. The *nuclear magneton*, μ_n is obtained by replacing the mass of the electron by the mass of the proton and is, therefore given by

$$\mu_n = \mu_b \, m_e/m_p = 5.05 \times 10^{-27} \text{ Am}^2.$$

Manometer. A device for measuring pressure. A simple type is a U-shaped glass tube containing mercury or other liquid. The pressure differences between the arms of the tube is indicated by the difference in heights of the liquid.

Mass. Symbol : m. A measure of the quantity of matter in an object. Mass is determined in two ways : the *inertial mass* of a body determines its tendency to resist change in motion ; the *gravitational mass* determines its gravitational attraction for other masses. The SI unit of mass is the kilogram.

Martensite. Supersaturated solution of C in α-Fe. The principal constituent of hardened steels, the actual hardness being dependent on the carbon content and increasing as the carbon content increases.

Martensitic Transformation. Similar structures to martensite have now been found in other alloy systems, *e.g.* Cu—Al. Transformations of this type involving lattice shear and no diffusion are known as martensitic transformations.

Mass Action, Law of. Guldberg and Waage (1864) stated that the rate at which a substance reacts is proportional to its active mass (concentration), and hence the velocity of a chemical reaction is proportional to the products of the concentrations of the reactants. Thus, in the reaction

$$A+B \rightleftharpoons C+D$$

the velocity of the forward reaction is

$$V_f \parallel k_1[A][B]$$

where k_1 is a constant, and the square brackets denote concentration. Similarly, for the backward reaction,

$$V_b=k_2[C][D]$$

At equilibrium, $V_f=V_b$

or $$k_1[A][B]=k_2[C][D],$$

hence $$\frac{k_1}{k_2}=\frac{[C][D]}{[A][B]}=K$$

The constants k_1 and k_2 are called velocity constants, and K is called the equilibrium constant.

Mass Defect. That part of the mass of the constituent particles of a nucleus which is apparent as binding energy rather than as mass.

Mass Concentration. See concentration.

Mass-energy Equation. The equation $E=mc^2$, where E is the total energy (rest-mass energy + kinetic energy + potential energy) of a mass m, c being the velocity of light. The equation is a consequence of Einstein's Special theory of relativity ; mass is a form of energy and energy also has mass. Conversion of rest-mass energy into kinetic energy is the source of power in radioactive substances and the basis of nuclear-power generation.

Mass Number. The total number of particles (protons and neutrons) in the nucleus.

Mass Spectrograph. This is a refined modification of the parabola method of positive-ray analysis, by means of which the masses of ions were compared to an accuracy of about 1 part in 20000. By its aid Aston provided most of the early information concerning isotopes.

In the modern mass-spectrometer the relative abundance of isotopes or other ionized species is determined by measuring the positive ion currents arriving at a fixed electrometer with various controlled magnetic fields and accelerating potentials. Resolution may be to 1 part in 50,000 and accuracy to better than 1 part in 10^6.

Mass Spectrometer. An instrument for producing ions in a gas and analysing them according to their charge/mass ratio. The earliest experiments by J.J. Thomson used a stream of positive ions from a discharge tube, which were deflected by parallel electric and magnetic fields at right angles to the beam. Each type of ion formed a parabolic trace on a photographic plate (a *mass spectrograph*).

In modern instruments, the ions are produced by ionizing the gas with electrons. The positive ions are accelerated out of this ion source into a high-vacuum region. Here, the stream of ions is deflected and focussed by a combination of electric and magnetic fields, which can be varied so that different types of ion fall on a detector. In this way, the ions can be analysed according to their mass, giving a *mass spectrum* of the material. Mass spectrometers are used for accurate measurements of relative atomic mass for the analysis of isotope abundance. They can also be used to identify compounds and analyse mixtures. An organic compound bombarded with electrons forms a number of fragment ions. (Ethane (C_2H_6), for instance, might form CH^+, $C_2C_5^+$, CH_2^+, etc.). The relative proportions of different types of ions is used to find the structure of new compounds. The characteristic spectrum can also identify compounds by comparison with standard spectra.

Mass Spectrum. The results obtained from a mass spectrometer are called mass spectra. They consist of a series of lines of varying intensity at different (*m/e*) ratios. For elements, the individual lines correspond to different isotopes. For compounds individual lines may be due to the presence of isotopic species, fragmentation of the molecule or ionization of the parent species in the mass spectrograph.

Matrix. A continuous solid phase in which particles of a different solid phase are embedded.

Matrix Isolation. The formation and study of unstable species by reaction between two or more substrates generally in the atomic or small molecular form which are produced in a Knudsen cell and in which the reactive species are condensed out on a cold surface in the presence of a great excess of inert material (generally argon). Species such as metal clusters, actinide carbonyls, (M plus CO), PdC_6H_5 and uncomplexed $MgCl_2$ have been studied. Characterization is generally by spectroscopic methods.

Maxwell Symbol : Mx. A unit of magnetic flux used in the c.g.s. system. It is equal to 10^{-8} Wb.

Maxwell-Boltzmann Distribution. A law describing the distribution of speeds among the molecules of a gas. In a system consisting of N molecules that are independent of each other except that they exchange energy on collision, it is clearly impossible to say what velocity any particular molecule will have. However, statistical statements regarding certain functions of the molecules were worked out by James Clerk Maxwell (1831-79) and Ludwig Boltzmann (1844-1906). One form of their law states that $n = N\exp(-E/RT)$, where n is the number of molecules with energy in excess of E. T is the thermodynamic temperature, and R is the gas constant.

Mcleod Gauge. A vacuum gauge in which a relatively large volume of a low-pressure gas is compressed to a small volume in a glass apparatus. The volume is reduced to an extent that causes the pressure to rise sufficiently to support a column of fluid high enough to read. This simple device, which relies on Boyle's law, is suitable for measuring pressures in the range 10^3 to 10^{-3} pascal.

Mean Free Path. The average distance traversed by a molecule between collisions with another molecule. At s.t.p. the mean path for hydrogen molecules is 1.8×10^{-6} cm, for nitrogen free 9.5×10^{-6} cm and for carbon dioxide 6.3×10^{-6} cm.

Mean Free Time. The average time that elapses between the collisions of the molecules in a gas, the electrons in a crystal, the neutrons in a moderator, etc. See mean free path

Mechanism of Reaction. The manner in which a reaction occurs. It is expressed in terms of a series of chemical equations. Thus the mechanism of the reaction of hydrogen peroxide with hydriodic acid may be written as

$$HI+H_2O_2 \longrightarrow H_2O+HOI$$

$$HOI+HI \longrightarrow H_2O+I_2$$

Mega Symbol : M. A prefix denoting 10^6. For example, 1 megahertz (MHz)$=10^6$ hertz (Hz).

Membrane. A thin pliable sheet of tissue or other material acting as a boundary. The membrance may be either natural (as in cells, skin, etc.) or synthetic modifications of natural materials (cellulose derivatives or rubbers). In many physicochemical studies membranes are supported on porous materials, such as porcelain, to provide mechanical strength. Membranes are generally permeable to some degree.

Membrane Hydrolysis. When a colloidal electrolyte consisting, *e.g.*, of giant anions and normal cations, is separated by a membrance from pure water, the cations are able to diffuse out but the colloidal anions are retained. For each positive ion which penetrates the membrane, a molecule of water will dissociate to provide a compensating OH^- ion, the corresponding H^+ ion tending to diffuse through the membrane to replace the original cation of the colloidal electrolyte. The liquid inside the dialyser thus becomes acid, that outside the dialyser alkaline. This process is termed membrane hydrolysis. See *also* colloids, colloidal electrolyte, membrane equilibrium.

Mercury Cell. A voltaic or electrolytic cell in which one or both of the electrodes consists of mercury or an amalgam. Amalgam electrodes are used in the Daniell cell and the Weston cadmium cell. Flowing mercury electrodes are also used in electrolytic cells. In the production of chlorine, sodium chloride solution is electrolysed in a cell with carbon anodes and a flowing mercury cathode. At the cathode, sodium metal is formed, which forms an amalgam with the mercury.

Metal Fatigue. A cumulative effect causing a metal to fail after repeated applications of stress, none of which exceeds the ultimate tensile strength. The *fatigue strength* (or *fatigue limit*) is the stress that will cause failure after a specified number (usually 10^7) of cycles. The number of cycles required to produce failure decreases as the level of stress or strain increases. Other factors, such as corrosion, also reduce the fatigue life.

Metallic Bond. A bond formed between atoms of a metallic element in zero oxidation state and in an array of similar atoms. The outer electrons of each atom are regarded as contributing to an 'electron gas', which occupies the whole crystal of the metal. It is the attraction of the positive atomic cores for the negative electron gas that provides the strength of the metallic bond.

Quantum mechanical treatment of the electron gas restricts its energy to a series of 'bands'. It is the behaviour of electrons in this bands that gives rise to semiconductors.

Metallic Conduction. Whereas electrolytic conduction arises from the movement of ions and results in decomposition of the conductor, conduction in metals arises from the movement of electrons and does not cause decomposition. The conductivity of a metal may be explained by assuming that the 'sea' of electrons behaves as gas molecules observing the kinetic laws. Each atom in the metal contributes only one or two electrons to the energy state known as the 'conduction band ; these are the electrons which are free to move.

Metallic Crystal. A crystalline solid in which the atoms are held together by metallic bonds. Metallic crystals are found in some interstitial compounds as well as in metals and alloys.

Metallography. The study of alloys and metals by various physical methods in order to elucidate constitution and structure.

Metalloid (Semimetal). Any of a class of chemical elements intermediate in properties between metals and non-metals. The classification is not clear cut, but typical metalloids are boron, silicon, germanium, arsenic, and tellurium. They are electrical semiconductors and their oxides are amphoteric.

Metastable State. Refers to a condition of a system or body in which it appears to be in stable equilibrium but, if disturbed, can settle into a lower energy state. For example, supercooled water is liquid below 0°C (at standard pressure). When a small crystal of ice or dust (for example) is introduced, rapid freezing occurs.

Metastable State. A condition of a system in which it has a precarious stability that can easily be disturbed. It is unlike a state of stable equilibrium in that a minor disturbance will cause a system in a metastable state to fall to a lower energy level. A book lying on a table is in a state of stable equilibrium a thin book standing on edge is in metastable equilibrium. Supercooled water is also in a metastable state. It is liquid below 0°C ; a grain of dust or ice introduced into it will cause it to freeze. An excited state of an atom or nucleus that has an appreciable lifetime is also metastable.

Metre Symbol m. The SI unit of length, being the length of the path travelled by light in vacuum during a time interval of $1/(2.99\ 792\ 458 \times 10^8)$ second. This definition, adopted by the General Conference on Weights and Measures in October, 1983, replaced the 1967 definition based on the krypton lamp, *i.e.* 1650763.73 wavelengths in a vacuum of the radiation corresponding to the transition between the levels $2p^{10}$ and $5d^5$ of the nuclide krypton-86. This definition (in 1958) replaced the older definition of a metre based on a platinum iridium bar of standard length. When the metric system was introduced in 1791 in France, the metre was intended to be one ten-millionth of the earth's meridian quadrant passing through Paris. However, the original geodetic surveys proved the impractibility of such a standard and the original platinum metre bar, the *metre des archives*, was constructed in 1793.

Metric System. A decimal system of units originally devised by a committee of the French Academy, which included J.L. Lagrange and P.S. Laplace, in 1791. It was based on the metre, the gram defined in terms of the mass of a cubic centimetre of water, and the second. This centimetre-gram-second system (see c.g.s. units) later gave way for scientific work to the metre-kilogram-second system (see m.k.s. units) on which SI units are based.

Mho. See siemevs.

Micelle. A term used to describe the colloidal particles in aggregation colloids, *e.g.* soaps and dyes. A micelle is a submicroscopic aggregate of molecules. Occasionally, the term micelle is usedfor polyelectrolyte particles.

Michaelis Constant. An experimentally determined parameter inversely indicative of the affinity of an enzyme for its substrate. For a constant enzyme concentration, the Michaelis constant is that substrate concentration at which the rate of reaction is half its maximum rate. In general, the Michaelis constant is equivalent to the dissociation constant of the enzyme-substrate complex.

Michaelis Reaction. A name originally applied in a restricted sense to the base catalysed addition of active methylene compounds (*e.g.*, malonic ester) to activated unsaturated systems (*e.g*, α : β unsaturated ketones). It is now used in a broader sense to define the conjugate additions of nucleophiles across double bonds conjugated with electron-withdrawing groups such as carbonyl (in $CH_3CH{=}CHCOCH_3$), nitrile ($CH_2{=}CHCN$) (see cyanoethylation), nitro, etc.

Micro Symbol : μ. A prefix used in the metric system to denote one-millionth. For example, 10^{-6} metre=1 micrometre (μm).

Microbalance. A sensitive balance capable of weighing masses of the order 10^{-6} to 10^{-9} kg.

Micron Symbol. μm A unit of length equal to 10^{-6} metre.

Microstate. In any system of particles the total energy of the system may be distributed among the particles in any one of a large number of equally probable arrangements. Each of these possible arrangements is termed a microstate.

Microwaves. A form of electromagnetic radiation, ranging in wavelength from about 1 mm (where it merges with infrared) to about 120 mm (bordering on radio waves). Microwaves are produced by various electronic devices including the klyston ; they are often carried over short distances in tubes of rectangular section called *wave guides*.

Spectra in the microwave region can give information on the rotational energy levels of certain molecules.

See *also* electromagnetic radiation.

Microwave Spectroscopy. Microwaves are electromagnetic waves with wavelengths in the range $10^{-2}-1$ m. Spectra are obtained by passing a microwave beam, produced by an oscillator using a Klystron electronic valve, through the gaseous sample ; the transmitted beam being detected by a crystal receiver. The resolution of such a microwave spectrometer is vastly superior to the best infra-red instruments and wavelengths are measured to seven figures. The spectra correspond to transitions between levels very close in energy, and information is derived about pure rotational spectra. Microwave spectroscopy is being used for the identification of molecules in the interstallar medium.

Migration 1. The movement of a group, atom, or double bond from one part of a molecule to another. 2. The movement of ions under the influence of an electric field.

Milli Symbol. m. A prefix denoting 10^{-3}. For example, 1 millimetre (mm)$=10^{-3}$ metre (m).

Millimetre of Mercury. See mmHg.

Miscible. Denoting combinations of substances that, when mixed, give rise to only one phase; *i.e.*, substances that dissolve in each other.

Miscibility. A term used to denote the extent of mixing. Gases mix in all proportions and are said to be completely miscible. However, liquids may be completely miscible, *e.g.*, alcohol and water, only partially miscible, *i.e.*, will only dissolve in each other to a limited extent, *e.g.*, aniline and water ; almost completely immiscible, *e.g*, water and mercury. The miscibility of liquids depends upon their chemical and physical similarities, in particular on their internal pressures. Liquids which have approximately equal internal pressures are miscible in all proportions.

Mitscherlich's Law (Law of Isomorphism). Substances that have the same crystal structure have similar chemical formulae. The law can be used to determine the formula of an unknown compound if it is isomorphous with a compound of known formula.

Mixed Crystals. Crystals deposited together from a solution containing to substances (which are normally isomorphous). Also termed solid solutions.

Mixture. Two or more substances forming a system in which there is no chemical bonding between the two. In homogeneous mixtures (*e.g.*, solutions or mixtures of gases) the molecules of the substances are mixed, and there is only one phase. In heterogeneous mixtures (*e.g.*, gunpowder or certain alloys) different phases can be distinguished. Mixtures differ from chemical compounds in that :

(1) The chemical properties of the components of a mixture are the same as those of the pure substances.

(2) The mixture can be separated by physical means (*e g.*, distillation or crystallization) or mechanically.

(3) The proportions of the components can vary. Some mixtures (*e.g* , certain solutions) can only vary in proportions between definite limits.

M.K.S. Units. A metric system of units devised by A. Giorgi (and sometimes known as *Giorgi units*) in 1901. It is based on the metre, kilograms and second and grew from the earlier c g.s. units. The electrical unit chosen to augment these three basic units was the ampere and the permeability of space (magnetic constant) was taken as 10^{-7} Hm^{-1}. To simplify electromagnetic calculations the magnetic constant was later changed to $4\pi \times 10^{-1}$ Hm^{-1} to give the *rationalized* MKSA *system.* This system, with some modifications, formed the basis of SI units now used in all scientific work.

MmHg. A unit of pressure equal to that exerted under standard gravity by a height of one millimetre of mercury, or 133.322 pascals.

Mobility, Ionic. According to Kohlrausch the equivalent conductivity at infinite dilution (Λ_∞) of a salt can be expressed as the sum of he equivalent ionic conductivities of the cation (λ_+) and anion (λ_-). The absolute velocities of migration of the ions, the ionic mobilities (u), which are the mean velocities of the ions in cm per sec for a potential gradient of 1 volt per cm are given by

$$u_+ = \frac{\lambda_+}{F} \text{ and } u_- = \frac{\lambda_-}{F}$$

where F is the Faraday.

Molal Concentration. See concentration.

Molality. The concentration of a solution 'expressed in terms of the number of moles of solute dissolved in 1 litre or solvent. Not to be confused with molarity.

Molar. Denoting that an extensive physical property is being expressed per amount of substance, usually per mole. For example, the molar heat capacity of a compound is the heat capacity of that compound per unit amount of substance ; in SI units it would be expressed in JK0[1] mol[1]0.

Molar Conductivity Symbol : A. The conductivity of that volume of an electrolyte that contains one mole of solution between electrodes placed one metre apart.

Molar Heat. The amount of heat required to raise the temperature of 1 mole of a substance by 1°C, at either constant pressure or constant volume.

Molar Heat Capacity. See heat capacity.

Molarity. A measure of the concentration of solutions based upon the number of molecules or ions present, rather than on the mass of solute, in any particular volume of the solution. The molarity (M) is the number of moles of solute in one cubic decimetre (litre). Thus a 0.5M solution of hydrochloric acid contains $0.5 \times (1+35.5)$g HCl per dm^3 of solution.

Molar Volume (Molecular Volume). The volume occupied by a substance per unit amount of substance.

Molarity. The concentration of a solution expressed as the number of moles of the solute in 1 litre of solution.

Mole. The quantity of a substance which contains one gram formula weight of the substance. One mole of any substance contains 6.023×10^{23} (Avagadro's number) molecules or atoms. Formerly called one gram molecule.

Molecular Beam. A beam of molecules in which all the molecules have velocities, and therefore energies, which fall into a very narrow range. Such beams may be used to measure molecular velocity distributions, molecular collision cross-sections and in kinetic studies.

Molecular Diameters. Althogh very few molecule, are in fact spherical, for the purpose of chemical kinetics they are assumed to be so. Some typical values are : H_2, 2.38×10^{-8} cm ; O_2, 3.19×10^{-8} cm ; NH_3, 3.9×10^{-2} cm ; and C_6H_6, 6.6×10^{-8} cm.

Molecular Distillation. Distillation in high vacuum (about 0.1 pascal) with the condensing surface so close to the surface of the evaporating liquid that the molecules of the liquid travel to the condensing surface without collisons. This technique enables very much lower temperatures to be used than are used with distillation at atmospheric pressure and therefore, heat-sensitive substances can be distilled. Oxidation of the distillate is also eliminated as there is no oxygen present.

Molecularity. The total number of reacting molecules in the individual steps of a chemical reaction. Thus, a unimolecular step has molecularity 1, a bimolecular step 2, etc. Molecularity is always an integer, whereas the order of a reaction need not necessarily be so. The molecularity of a reaction gives no information about the mechanism by which it takes place.

Molecular Sieve. Porous crystalline substances, especially aluminosilicates (see zeolite), that can be dehydrated with little change in crystal structure. As they form regularly spaced cavities, they provide a high surface area for the adsorption of smaller molecules. The general formula of these substances is $M_nO.Al_2O_3.xSiO_2.yH_2O$, where M is a metal ion and n is twice the reciprocal of its valency. Molecular sieves are used as trying agents and in the separation and purification of fluids. They can also be loaded with chemical substances, which remain separated from any reaction that is taking place around them, until they are released by heating or by displacement with a more strongly adsorbed substance. They can thus be used as cation exchange mediums and as catalysts and catalyst supports.

Molecular Spectrum. Band spectrum and continuous spectrum.

Molecule, Mass Of. The actual mass of a molecule can be obtained by dividing the gram molecular weight by Avagadro's number. Thus the mass of a nitrogen molecule is

$$\frac{28}{6\ 02 \times 10^{23}} = 4.65 \times 10^{23} \text{ g}$$

Mole Fraction. The mole fraction (X_a) of a compound A in a mixture containing in addition only compounds B, C and D is the ratio of the number of molecules of A to the total number of molecules of A+B+C+D. In practice, the concentrations of A, B, C and D are expressed as fractions of the number of moles of each present. The sum of the mole fractions of the components of the system, *i.e.* $X_A+X_B+X_C+X_D$ is equal to unity.

Monatomic. Denoting a molecule, radical, or ion consisting of only one atom. For example, helium is a monatomic gas and H is a monatomic radical.

Monochromatic Radiation. Radiation of a single wavelength. See monochromator.

Monochromator. A device, prism, grating, single crystal for selecting a single wavelength from a wide wavelength range of radiation or particles. Used in spectrophotometers and in X-ray diffraction equipment.

Monoclinic System. The crystal system with one 2-fold axis and/or one plane of symmetry. A unit cell has three unequal axes one perpendicular to the other two which are obliquely inclined. *e.g.* Na_2CO_3, $10H_2O$.

Moseley's Law. Lines in the X-ray spectra of elements have frequencies that depend on the proton number of the element. For a set of elements, a graph of the square root of the frequency of X-ray emission against proton number is a straight line (for spectral lines corresponding to the same transition).

Mossbauer Effect. The resonance fluorescence by γ-radiation of an atomic nucleus, returning from an excited state to the ground state. The resonance energy is characteristic of the chemical environment of the nucleus and Mossbauer spectroscopy may be used to yield information about this chemical environment, *e.g.*, the state of oxidation of iron in meteorities has been determined by Mossbauer spectroscopy. Used particularly in the study of Fe, Sn and Sb compounds.

Mulliken Symbols. The designators, arising from group theory, of the electronic states of an ion in a crystal field. A and B are singly degenerate, E doubly degenerate, T triply degenerate states. Thus a D state of a free ion shows E and T_2 states in an octahedral field.

Multiplet. In a spectrum the individual lines often consist of two or more closely grouped fine lines, which make up the fine structure of the original line. Each such group, which constitutes a line in the ordinary spectrum, is called a multiplet. In an electronic spectrum this fine structure arises from the slightly differing amounts of energy an electron possesses due to its spin. The multiplicity is the maximum possible number of such energy values which an atom in a specified state can possess by virtue of the electron spin. A multiplet which arises from the electronic transition where there are two spin states is called a *doublet* ; where there are three spin states a *triplet* is found.

N

Nano. Symbol *n*. A prefix used in the metric system to denote 10^{-9}. For example, 10^{-9} second=1 nanosecond (ns).

Negative Adsorption. When adsorption at a surface occurs from a solution the concentration of some components may be higher at the surface than in the bulk, whilst with other components the concentration will necessarily be less at the surface than in the bulk. These latter components are said to undergo negative adsorption. According to the Gibbs's adsorption equation any solute which increases the interfacial tension of the system will be negatively adsorbed, *e.g.*, the negative adsorption of sodium chloride at the air/water interface, which has been confirmed experimentally.

Nematic Liquid Crystals. See liquid crystals.

Nephelauxetic Effect. The changes in the electronic ($d{\rightarrow}d$, etc.) spectra of complexes due to varying degrees of covalent character in the metal-ligand bonds. Measured as the ratio, β, of the Racah parameter B in the complex as compared with B in the free ion. The greater β the greater the ionic character in the bond.

Nerast Equation. This equation relates the e.m.f. of a cell to the concentrations or, more accurately, the activities of the reactants and products of the cell reaction. For a reaction

$$aA+bB\rightleftharpoons cC+dD$$

the e.m.f., E of the cell is given by

$$E=E^{\circ}-RT\ln\left(\frac{a_C{}^c a_D{}^d}{a_A{}^a a_B{}^b}\right)$$

where E° is the standard e.m.f. of the cell and the a_n are the activities of the reactants and products under a given set of conditions.

Nernst Heat Theorem. A statement of the third law of thermodynamics in a restricted form : if a chemical change takes place between pure crystalline solids at absolute zero there is no change of entropy.

Neutral. Describing a compound or solution that is neither acidic nor basic. A neutral solution is one that contains equal numbers of both protonated and deprotonated forms of the solvent.

Neutralization. The stoichiometric reaction of an acid and a base in volumetric analysis. The neutralization point or end point is detected with indicators.

Neutralization, Heat of. The amount heat evolved when 1 g equivalent of an acid is neatralized by 1 g equivalent of a base. For strong acids and strong bases in dilute solution the only reaction which occurs is $H^+ + OH^- \rightarrow H_2O$ and the heat of neutralization has a constant value of 57.35 kJ.

Neutralization Value. The acid number or base number of a petroleum of other material. Total acid number is given by the amount of base, expressed in mg of KOH required to neutralize all acid constituents present in 1 g of sample. Total base number is the quantity of acid, expressed in terms of the equivalent number of mg of KOH required to neutralize all the basic constituents present in 1 g of sample.

Colorimetric and potentiometric methods are frequently used and are specified for particular materials and products.

Neutron Number. Symbol N. The number of neutrons in an atomic nucleus of a particular nuclide. It is equal to the difference between the nucleon number and the atomic number.

Newton. Symbol : N. The SI unit of force, equal to the force needed to accelerate one kilogram by one metre second^{-2}. 1 N=1 kg m s^{-2}.

Nickel-Iron Accumulator. (Edison cell, Nife or NIFE cell). A type of secondary bell in which the electrodes are formed by steel grids. The positive electrode is impregnated with a nickel-nickel hydride mixture and the negative electrode is impregnated with iron oxide. Potassium hydroxide solution forms the electrolyte. The nickel-iron cell is lighter and more durable than the lead accumulator, and it can work with higher currents. Its e.m.f. is approximately 1.3 volts.

The reaction on discharge is

$$2NiOOH.H_2O + Fe \rightarrow 2Ni(OH)_2 + Fe(OH)_2,$$

the reverse occurring during chaging. Each cell gives an e.m.f. of about 1.2 volts and produces about 100 kJ per kilogram during each discharge. *Compare* lead-acid accumulator.

NIFE Cell. See nickel-iron accumulator.

N.M.R. Shift Reagents. Various lanthanide chelates principally of β-diketones are useful for simplifying n.m.r. spectra because of their ability to cause spectacular changes in chemical shifts of nuclei adjacent to an electronegative substituent in a molecule Europium complexes, added as solids or solutions, cause downfield shifts and praesodymium complexes result in upfield changes in resonances.

Non-Aqueous Titrations. Analytical procedures carried out in organic and other non-aqueous media because of insolubility of the reactants in water, or because some acids and bases cannot be titrated in water. Thus H_2SO_4 and $HClO_4$ can be titrated as a mixture in methylisobutyl ketone whilst amines can be titrated in 3-methylsulpholane.

Normal Liquid. A liquid in which the molecules have no tendency to associate. The simpler liquid hydrocarbons approximate to this type, while water, ethanoic acid and alcohol are examples of abnormal liquids.

NQR. Nuclear quadruple resonance.

NTP. Normal temperature (298 K) and pressuse (760 mm).

Nuclear Magnetic Resonance (NMR). A method of investigating nuclear spin. In an external magnetic field the nucleus can have certain quantized energy states, corresponding to certain orientations of the spin magnetic moment. Hydrogen nuclei, for instance, can have two energy states, and transitions between the two occur by absorption of radiofrequency radiation. In chemistry, this is the basis of a spectroscopic technique for investigating the structure of molecules. Radio frequency radiation is fed to a sample and the magnetic field is changed slowly. Absorption of the radiation is detected when the difference between the nuclear levels corresponds to absorption of a quantum of radiation. This difference depends slightly on the electrons around the nucleus *i.e.*, the position of the atom in the molecule. Thus a different absorption frequency is seen for each type of hydrogen atom. In ethanol, for example, there are there frequencies, corresponding to hydrogen atoms on the CH_3, the CH_2, and the OH. The intensity of absorption also depends on the number of hydrogen atoms (3 : 2 : 1). NMR spectroscopy is a powerful method of finding the structures of organic compounds.

Information from an n.m.r. spectrum is classified into the chemical shift, δ (the relative shift from a standard [Me_4Si for ^1H, CCl_3F for ^{19}F] which is rendered independent of the field), and the conpling coustants, J, which are determined directly from the spectra.

Complex n.m.r. spectra can bc simplified by the use of n.m.r. shift reagents.

Nuclear Paramagnetism. The nuclei of many species have spin and the total angular momentum is given by

$$\frac{h}{2\pi}\sqrt{I(I+1)}$$

where I is the nuclear spin number (I=0, $\frac{1}{2}$, 1, $\frac{3}{2}$ ...). I=0

corresponds to a nucleus without spin. Since nuclei are also charged they behave as magnetic dipoles and nuclear magnetic resonance techniques can be used to study nuclei with $I>0$.

Isotope	*Spin no. (I)*	*% abundance*
^{1}H	½	99.98
^{2}H	1	0.0156
^{12}C	0	98.9
^{13}C	½	1.1
^{14}N	1	99.62
^{15}N	½	0.38
^{19}F	½	100
^{31}P	½	100

Nuclear Quadrupole Moment. The electric quadrupole moment of the nucleus. Affected by chemical environment and can be used to investigate structure (NQR spectroscopy).

Naclear Spin. From the hyperfine structure of certain spectra (see multiplet) it has been concluded that the nucleus may have a property with the characteristics of spin. The angular momentum of nuclear particles is expressed in units of $(h/2\pi)$. (h=Planck's constant). See nuclear magnetic resonance.

Nucleus. The central core of an atom that contains most of its mass. It is positively charged and consists of one or more nucleons (protons of neutrons). The positive charge of the nucleus is determined by the number of protons it contain (see atomic number) and in the neutral atom this is balanced by an equal number of electrons, which move around the nucleus.

Nuclide. A species of atom characterized by the constitution of its nucleus. Commonly used in the context of radioactive decay. More than 1600 nuclides are known.

Nucleon Number (Mass Number) Symbol : A. The number of nucleons in an atomic nucleus of a particular nuclide.

Nujol. A trade name for a heavy medicinal liquid paraffin. Extensively used as a mulling agent in spectroscopy.

O

Occlusion. A term used to denote either the retention of a gas or of solids by a metal, or the absorption of an electrolyte by a precipitate. Generally used to indicate the penetration of the metal lattice by atoms or molecules of the occluded substance, sometimes with the formation of an interstitial compound.

Oersted. Symbol : Oe. A unit of magnetic field strength in the c.g.s. system. It is equal to $10^3/4\pi$ A m^{-1}.

Ohm. Symbol : Ω. It is the SI unit of electrical resistance which is equal to a resistance that passes a current of one ampere when there is an electric potential difference of one volt across it. 1 $\Omega = 1$ V A^{-1}. Formerly, it was defined in terms of the resistance of a column of mercury under specified conditions.

Onium Ion. An ion which is formed by addition of a proton (H^+) to a molecule. The hydronium ion (H_3O^+) and ammonium ion (NH_4^+) are examples.

Optical Glass. Glass used in the manufacture of lenses, prisms, and other optical parts. It must be homogeneous and free from bubbles and strain. Optical *crown glass* may contain potassium or barium in place of the sodium of ordinary crown glass and has a refractive index in the range 1.51 to 1.54. ***Flint glass*** contains lead oxide and has a refractive index between 1.58 and 1.72. Higher refractive indexes are obtained by adding lanthanoid oxides to glasses ; these are now known as lanthanum crowns and flints.

Optical Rotatory Dispersion, ORD. A description of the changes in optical rotation of an optically active molecule, organic or inorganic, with the wavelength of the light. The variation of specific rotation with wavelength increases as an absorption band is approached, and the rotation may pass through a maximum or minimum when the absorption band is crossed. This is the so-called Cotton-effect. The effect is particularly sensitive to structural features of many molecules and is used in structural and stereochemical studies. ORD may be induced by a magnetic field.

Optoacoustic Spectrometry. A spectrophotometric technique whereby modulated electromagnetic radiation falling upon the sample causes heating when absorption occurs. The heating causes a pressure change in the cell containing the sample which is detected by a microphone. Potentially can be used to examine most regions of the electromagnetic spectrum.

Optical Rotation. Rotation of plane-polarized light. See optical activity.

Orbital Angular Momentum, L. The total angular momentum of a set of electrons in an atom or ion. For filled and empty shells, L is zero.

Orbital Moment. The component of the magnetic moment of an atom or ion resulting from the motion of the electron around the nucleus.

Order of Reaction. The power to which the concentration terms are raised in the mathematical equation which describes the variation of the rate of a reaction with concentration of reactants at a given temperature.

For example, in the expression

$$\text{rate}=k[\text{A}]^x[\text{B}]^y$$

x is called the order with respect to A, y the order with respect to B, and $(x+y)$ the order overall. The values of x and y are not necessarily equal to the coefficients of A and B in the molecular equation. Order is an experimentally determined quantity derived without reference to any equation or mechanism. Fractional orders do occur. For example, in the reaction

$$CH_3CHO \rightarrow CH_4+CO$$

the rate is proportional to $[CH_3CHO]^{1.5}$, *i.e.*, it is of order 1.5.

Orifice Meter. An important instrument extensively used for measuring the flow of liquids and gases in pipes.

Orifice meters are widely used because they are simple and cheap to manufacture and install, and their design is so well standardized they do not require calibration.

Orthorhombic System. The crystal system with a 2-fold axis at the intersection of two planes of symmetry or perpendicular to other 2-fold axes. A unit cell has three unequal axes at right angles. *E.g.*, KNO_2.

Osmometer. See osmosis.

Osmosis. When a solvent and a solution are separated by a semi-permeable membrane, such as a piece of parchment or a plant or animal membrane, the solvent tends to flow spontaneously through the membrane into the solution. This process is known as osmosis. A similar flow is observed when two solutions of different chemical potential are separated by a semi-permeable membrane.

Osmotic Pressure. The excess hydrostatic pressure which must be applied in order to counter-balance the process of osmosis. Thus a solution under this excess pressure can come to osmotic equilibrium with the same solvent on the other side of a membrane permeable only to the solvent molecules. For dilute solutions of non-eleetrolytes the osmotic pressure is equal to pressure they would exert if they existed as gases under the same conditions of molecular concentration and temperature. For dilute solutions the osmotic pressure (π) is given by

$$\pi = \frac{cRT}{M}$$

where c is the concentration of solute, of molecular weight M, in grammes per unit volume. R is the gas constant and T, the absolute temperature.

Ostwald's Dilution Law. An expression for the degree of dissociation of a weak electrolyte. For example, if a weak acid dissociates in water

$$HA \rightleftharpoons H^+ + A^-$$

the dissociation constant K_a is given by

$$K_a = \alpha^2 n/(1-\alpha)V$$

where α is the degree of dissociation, n the initial amount of substance (before dissociation), and V the volume. If α is small compared with 1, then $\alpha^2 = KV/n$; *i.e.*, the degree of dissociation is proportional to the square root of the dilution.

The law was first put forward by W. Ostwald (1853-1932) to account for electrical conductivities of electrolyte solutions.

Over-Voltage, Over-Potential. The excess potential, over and above that of the reversible electrode, that must be applied to an electrode for electrolysis to proceed at a measurable rate. *E.g*, at a platinized platinum electrode the evolution of hydrogen will occur practically at the reversible electrode potential. However, at a mercury electrode no hydrogen is evolved until an over-voltage of approximately 1 volt is reached. Over-voltage is of great practical importance in processes such as electro-plating, electro-analysis, electro-reduction, polarography, etc., although as yet the theory of the phenomena is incomplete. See hydrogen over-voltage.

Oxygen Cathode. A platinum or gold elecIrode used in amperometry for measuring oxygen content by measuring the rate of cathodic reduction of O_2. The cathode is used with a reference electrode. A Clark electrode is a typical oxygen electrode.

P

Paramagnetism. If when a substance is placed in a magnetic field it causes a greater concentration of the lines of magnetic force within itself than in the surrounding magnetic field, it is said to exhibit paramagnetism. Paramagnetism is associated with the presence of unpaired electrons in an ion or molecule. Magnetic susceptibility measurements of paramagnetic substances have contributed significantly to the theory of valency and knowledge of structure.

Partial Pressure. In a mixture of gases, the contribution that one component makes to the total pressure. It is the pressure that the gas would have if it alone were present in the same volume. See Dalton's law of partial pressures.

Partition. If a substance is in contact with two different phases then, in general, it will have a different affinity for each phase.

Part of the substance will be absorbed or dissolved by one and part by the other, the relative amounts depending on the relative affinities. The substance is said to be *partitioned* between the two phases. For example, if two immiscible liquids are taken and a third compound is shaken up with them, then an equilibrium is reached in which the concentration in one solvent differs from that in the other. The ratio of the concentrations is the *partition coefficient* of the system. The *partition law* states that this ratio is a constant for given liquids.

Partition Coefficient. If a solute dissolves in two non-miscible liquids, the partition coefficient is the ratio of the concentration in one liquid to the concentration in the other liquid.

Partition Law. In a heterogeneous system of two or more phases in equilibrium, the ratio of the activities (or less accurately the concentrations) of the same molecular species in the phases is a constant at constant temperature. This constant is termed the partition coefficient. The law is also known as the distribution law.

Patterson Function. The fourier transform of observed intensities after diffraction. From the Patterson map it is possible to determine the positions of scattering centres (atoms, electrons).

Pascal. Symbol : Pa. The SI unit of pressure, equal to a pressure of one newton per square metre ($1 \text{ Pa} = 1 \text{ N m}^{-2}$).

Passive. Describing a metal that is unreactive because of the formation of a thin protective layer. Iron, for example, does not dissolve in concentrated nitric (V) acid because of the formation of a thin oxide layer.

Penetration. A method of viscosity measurement used for solids or very viscous liquids, mainly bitumen, grease or wax The penetration of the distance in tenths of a millimetre that a standard needle or cone will penetrate under standard conditions of load, time and temperature.

Perfect Gas. See gas laws, kinetic theory.

Perfect Solution. See Raoult's law.

Permanent Gas. A gas, such as oxygen or nitrogen, that was formerly thought to be impossible to liquefy. A permanent gas is now regarded as one that cannot be liquefied by pressure alone at normal temperatures (*i.e.*, a gas that has a critical temperature below room temperature).

Peta-. Symbol ; P. A prefix used in the metric system to denote one thousand million, million times. For example, 10^{15} metres=1 petametre (Pm).

pH. The logarithm to base 10 of the reciprocal of the hydrogen-ion concentration of a solution. In pure water at 25°C the concentration of hydrogen ions is 1.00×10^{-7} moles l^{-1}, thus the pH equals 7 at neutrality. An increase in acidity increases the value of $[H^+]$, decreasing the value of the pH below 7. An increase in the concentration of hydroxide ion $[OH^-]$ proportionately decreases $[H^+]$, therefore increasing the value of the pH above 7 in basic solutions, pH values can be obtained approximately by using indicators. More precise measurements use electrode systems.

Phase. One of the physically separable parts of a chemical system. A system consisting of only one phase is said to be homogeneous. A system consisting of more than one phase is said to be heterogeneous.

E.g., in a closed system containing ice, water and water vapour, there are three phases. Since all gases are completely miscible with each other, they can only exhibit one phase. With liquids and solids several phases are possibie.

Phase Diagram. A graphical representation of the state in which a substance will occur at a given pressure and temperature.

The lines show the conditions under which more than one phase can coexist at equilibrium. For one-component systems (*e.g.*, water) the point at which all three phases can coexist at equilibrium is called the triple point and is the point on the graph at which the pressure-temperature curves intersect.

In one-component systems a pressure-temperature diagram is used ; in binary systems a three-dimensional diagram relating pressure, composition and temperature, or two-dimensional diagrams relating composition and temperature at constant pressure or composition and pressure at constant temperature are used.

Phase Equilibrium. A state in which the proportions of the various phases in a chemical system are fixed. When two or more phases are present at fixed temperature and pressure a dynamic condition will be established in which individual particles will leave one phase and enter another ; at equilibrium this will be balanced by an equal number of particles making the reverse change, leaving the total composition unchanged.

Phase Rule. First derived by Willard Gibbs (1876), the phase rule relates the number of phases, P, the number of degrees of freedom, F, and the number of components, C, in a chemical system according to the equation

$$P+F=C+2$$

Phase Space. See statistical mechanics.

pH Meter. An instrument used for measuring the pH of solutions, suspensions, etc., generally with the aid of a glass electrode. It consists essentially of an electrometer-valve potentiometer with an adjustment for backing off the asymmetry potential of the electrode. Both 'mill' type, which is the more accurate, and 'direct reading' instruments are available.

Phosphor. Refers to a substance that is capable of luminescence (including phosphorescence). Phosphors that release their energy after a short delay of between 10^{-10} and 10^{-4} second are sometimes called *scintillators*.

Phosphorescence. 1. The absorption of energy by atoms followed by emission of electromagnetic radiation. Phosphorescence is a type of luminescence, and is distinguished for fluorescence by the fact that the emitted radiation continues for some time after the source of excitation has been removed. In phosphorescence the excited atoms have relatively long lifetimes before they make transitions to lower energy states. However, there is no defined time distinguishing phosphorescence from fluorescence.

2. In general usage the term is applied to the emission of 'cold light'—light produced with a high temperature. The name comes from the fact that white phosphorus glows slightly in the dark as a result of a chemical reaction with oxygen The light comes from excited atoms produced directly in the reaction—not from the heat produced. It is thus an example of *chemiluminescence.* There are also a number of biochemical examples termed bioluminescence ; for example, phosphores-

cence is sometimes seen in the sea from marine organisms, or on rotting wood from certain fungi (known as 'fox fire').

Phot. A unit of illumination in the c.g.s. system, equal to an illumination of one lumen per square centimetre. It is equal to 10^4 lx.

Photochemical Reaction. A chemical reaction caused by light or ultraviolet radiation. The incident photons are absorbed by reactant molecules or free radicals, which undergo further reaction.

Photochemistry. The science dealing with chemical reactions brought about by the action of light. Only light which is absorbed by the system can produce chemical effects ; any light which passes through the material or is scattered is necessarily ineffective.

Photochromism. A change of colour occurring in certain substances when exposed to light. Photochromic materials are used in sunglasses that darken in bright sunlight.

Photoconductive Effect. See photoelectric effect.

Photodissociation. This is the process which occurs when a molecule on absorbing a quantum of light energy undergoes dissociation to smaller molecules, radicals or atoms. *E.g.*, propanone undergoes photodissociation to methyl radicals and carbon monoxide :

$$(CH_3)_2CO \rightarrow 2CH_3\cdot + CO$$

Photo-Electric Cells. Exposure of certain elements, particularly the alkali metals to light of suitable wavelength results in the emission of electrons. This phenomenon is known as the photo-electric effect. Since the intensity of emission is proportional to the intensity of the incident light, the effect can be made the basis of an instrument for measuring light intensity.

Photoelectric Effect. The emission of electrons from a solid (or liquid) surface when it is irradiated with electromagnetic radiation. For most materials the photoelectric effect occurs with ultraviolet radiation or radiation of shorter wavelength ; some show the effect with visible radiation.

In the photoelectric effect, the number of electrons emitted depends on the intensity of the radiation and not on its frequency. The kinetic energy of the electrons that are ejected depends on the frequency of the radiation. This was explained, by Einstein, by the idea that electromagnetic radiation consists of streams of photons. The ***photon energy*** is hv, where h is the Planck constant and v the frequency of the radiation. To remove an electron from the solid a certain minimum energy must be supplied, known as the ***work function***, ϕ. Thus, there is a certain minimum threshold frequency v_0 for radiation to eject electrons : $hv_0=\theta$. If the frequency is higher than this threshold the electrons are ejected. The maximum kinetic energy (W) of the electrons is given by ***Einstein's equation*** :

$$W=hv-\phi$$

The photoclectric effect also occurs with gases. See photo-ionization.

Photoelectron. An electron ejected from a solid, liquid, or gas by the photoelectric effect or by photoionization.

Photo-Electron Spectroscopy. Bombardment of an atom with photons of a specific energy results in the emission of an electron of energy E, where $E=(E_t-E_b)$, E_t and E_b being the respective energies of the photon and of binding of the ejected electron. The technique may be used for measuring binding energies of valence electrons or for inner core electrons. Inner core spectroscopy, more commonly termed ESCA (electron spectroscopy for chemical analysis) or X-ray photo-electron spectroscopy (XPS), generally uses Al($k\alpha$) 1486 eV as the photon source. Ultra-violet photo-electron spectroscopy using He^I, 21.2 eV and He^{II} 40.8 eV radiation is used for examining valence shell electrons.

Photoemission. The emission of photoelectrons by the photo-electric effect or by photoionization.

Photoionization. The ionization of atoms or molecules by electro-magnetic radiation. Photons absorbed by an atom may have sufficient photon energy to free an electron from its attraction by the nucleus. The process is

$$M+hv \rightarrow M^{+}+e^{-}$$

As in the photoelectric effect, the radiation must have a certain minimum threshold frequency. The energy of the photoelectrons ejected is given by $W=hv-I$, where I is the ionization potential of the atom or molecule.

Photoluminescence. A general term used to describe the emission of light as a result of an initial absorption of light. The phenomena of fluorescence and phosphorescence are examples of photoluminescence.

Photolysis. A chemical reaction that is produced by electromagnetic radiation (light or ultraviolet radiation). Many photolytic reactions involve the formation of free radicals.

Photon. A quantum of light energy. The energy of a photon is given by $E=hv$, where v is the frequency of the radiation and h in Planck's constant.

Photosensitization. A process in which a photo-chemical reaction is induced to occur by the presence of a substance (the photosensitizer) which absorbs the light but itself is substantially unchanged at the end of the reaction, the absorbed light energy being passed on to the main reactants. *E.g.*, when hydrogen is exposed to light of wavelength 2536 A, no absorption of the light occurs, and the hydrogen remains completely unaffected. If mercury vapour is added to the hydrogen, the mercury atoms are excited. When such an excited mercury atom collides with a hydrogen molecule, it can transfer some of its energy to the hydrogen, and cause it to dissociate into atoms :

$$Hg+H_2=Hg+2H$$

The hydrogen has apparently been made sensitive to the light which it does not absorb.

Photosensitive Substance. 1. Any substance that when exposed to electromagnetic radiation produces a photoconductive, photdelectric, or photovoltaic effect. 2. Any substance, such as the emulsion of a photographic film, in which electromagnetc radiation produces a chemical change.

Photostationary State In a chemical reaction system the term stationary state is used to indicate a state of equilibrium under specified conditions. In a photochemical reaction, a stationary state is attained when the rate of removal of reactants by light is equal to the rate of recombination or reaction of the products to yield the reactants. This condition is referred to as a photostationary state.

Photosynthesis. The process by which green plants build up their carbon compounds from atmospheric carbon dioxide using light as the energy source.

Phototropy. The reversible change of colour of certain substances when exposed to light of a certain wavelength.

pH Scale. A logarithmic scale for expressing the acidity or alkalinity of a solution. To a first approximation, the pH of a solution can be defined as $-\log_{10}c$, where c is the concentration of hydrogen ions in moles per cubic decimetre. A neutral solution at 25°C has a hydrogen-ion concentration of 10^{-7} mol dm^{-3}, so the pH is 7. A pH below 7 indicates an acid solution ; one above 7 indicates an alkaline solution. More accurately, the pH depends not on the concentration of hydrogen ions but on their activity, which cannot be measured experimentally. For practical purposes, the pH scale is defined by using a hydrogen electrode in the solution of interest as one half of a cell, with a reference electrode (*e.g.*, a calomel electrode) as the other half cell. The pH is then given by $(E-E_R)F/2.303RT$, where E is the e.m.f. of the cell and E_R the standard electrode potential of the reference electrode. In practice, a glass electrode is more convenient than a hydrogen electrode.

pH stands for 'potential of hydrogen'. The scale was introduced by S.P. Sorensen in 1909.

Physical Change. A change to a substance that does not alter its chemical properties. Physical changes (*e.g.*, melting, boiling, and dissolving) are comparatively easy to reverse.

Physical Chemistry. The branch of chemistry concerned with the effect of chemical structure on physical properties. It includes chemical thermodynamics and electrochemistry.

Physisorption. See adsorption.

Pico-. Symbol : p. A prefix used in the metric system to denote 10^{-12}. For example, 10^{-12} farad=1 picofarad (pF).

Piezoelectricity. The electric charge developed in anisotropic crystals when subjected to stress (*e.g.*, pressure). Used in electrical components and particularly in lighters.

pK. The logarithm to the base 10 of the reeiprocal of an acid's dissociation constant :

$$\log_{10}(1/K)$$

For any acid HA which partially dissociates in solution, the equilibrium

$HA(solv.) \rightleftharpoons H^+(solv.) + A^-(solv.)$ is defined by an equilibrium constant K, where

$$K = \frac{[H^+][A^-]}{[HA]}$$

The negative logarithm of the equilibrium constant is termed the pK, *i.e.*, pH=−log K. For acids, the pK is denoted by pK_a. Similarly for, a base BOH, the pK_b is given by $pK_b = -\log K_b$, where

$$K_b = \frac{[B^+][OH^-]}{[BOH]}$$

pK Value. A measure of the strength of an acid on a logarithmic scale. The pK value is given by $\log_{10}(1/K_a)$, where K_a is the acid dissociation constant. pK values are often used to compare the strengths of different acids.

Planck Constant. Symbol : *h*. A fundamental constant ; the ratio of the energy (W) carried by a photon to its frequency (v). A basic relationship in the quantum theory of radiation is $W = hv$. The value of *h* is 6.626196×10^{-34} J s. The Planck constant appears in many relationships in which some observable measurement is quantized (*i.e.*, can take only specific separate values rather than any of a range of values).

Plane Polarization. A type of polarization of electromagnetic radiation in which the vibrations take place entirely in one plane.

Plane-Polarized Light. See polarization of light.

Pleochroic. Denoting a crystal that appears to be of different colours, depending on the direction from which it is viewed. It is caused by polarization of light as it passes through an anisotropic medium.

Point Group. A set of symmetry elements passing through or referring to a single point. Refers to molecules—in lattices where symmetry elements may not all pass through one point the corresponding arrangement is referred to as a space-group.

Poise. A c.g.s. unit of viscosity equal to the tangential force in dynes per square centimetre required to maintain a difference in velocity of one centimetre per second between two parallel planes of a fluid separated by one centimetre. 1 poise is equal to 10^{-1} N s m^{-2}.

Poison. A substance that destroys catalyst activity.

Polariscope (Polarimeter). A device used to study optically active substances (see optical activity). The simplest type of instrument consists of a light source, collimator, polarizer, and analyser. The specimen is placed between polarizer and analyser, so that any rotation of the plane of polarization of the light can be assessed by turning the analyser.

Polarizability. When a molecule is placed in an electric field the electrons tend to be drawn away from the nuclei with the result that a dipole is induced in the molecule. The strength of the dipole divided by the field strength is termed the polarizability of the molecule. Such *electronic* polarizability can be determined from measurements of refractive index at wavelengths in the visible spectral region. Similar measurements in the far i.r. give, additionally, the *atomic* polarizability. Polarizability has units of volume.

Polarization. 1. The restriction of the vibrations in a transverse wave so that the vibraiton occurs in a single plane. Electromagnetic radiation, for instance, is a transverse wave motion. It can be thought of as an oscillating electric field and an oscillating magnetic field, both at right angles to the direction of propagation and at right angles to each other. Usually, the electric vector is considered since it is the electric field that interacts with charged particles of matter and causes the effects. In 'normal' unpolarized radiation, the electric field oscillates in all possible directions perpendicular to the wave direction. On reflection or on transmission through certain substances (*e.g.*, Polaroid) the field is confined to a single plane. The radiation is then said to be *plane-polarized*.

2. See polarizability.

3. The reduction of current in a voltaic cell, caused by the build-up of products of the chemical reaction. Commonly, the cause is the build-up of a layer of bubbles (*e.g.*, of hydrogen). This reduces the effective area of the electrode, causing an increase in the cell's internal resistance. It can also produce a back e.m.f. Often a substance such as manganese (IV) oxide (a *depolarizer*) is added to prevent hydrogen build-up.

Polarization of Light. The process of confining the vibrations of the electric vector of light waves to one direction. In unpolarized light the electric field vibrates in all directions perpendicular to the direction of propagation. After reflection or transmission through certain substances (see Polaroid) the electric field is confined to one direction and the radiation is said to be *plane-polarized light*. The plane of plane-polarized light can be rotated when it passes through certain substances (see optical activity.

In *circularly polarized light*, the tip of the electric vector describes a circular helix about the direction of propagation with a frequency equal to the frequency of the light. The magnitude of the vector remains constant. In *elliptically polarized light*, the vector also rotates about the direction of propagation but the amplitude changes ; a projection of the vector on a plane at right angles to the direction of propagation describes an ellipse. Circularly and elliptically polarized light are produced using a retardation plate.

Polar Molecule. As the two electrons in a covalent link are not shared equally (different electro-negativities). there is a separation of charge over the bond and, if the effects are not cancelled out, over the whole molecule. Such a molecule is known as a polar molecule. Thus in covalent hydrogen chloride

$\overset{\delta+}{H} - \overset{\delta-}{Cl}$ the chlorine has a greater share of the bonding pair than the hydrogen and the molecule has a resultant dipole moment. Lone pairs of electrons also make molecules strongly polar.

Polarography. A highly developed electrochemical method of analysis, particularly suitable for dilute solutions of substances which are susceptible to electrolytic reduction at a mercury cathode. A dropping-mercury electrode (DME) is used and the polarogram, the current-voltage curve, for the solution is recorded. This polarogram shows a step for each reducible species present in solution. This step is of characteristic potential and of height proportional to the concentration of the component. In favourable cases several components may be detected and determined quantitatively from the polarogram. The method is particularly useful for determining trace amounts of metal and it can be used for investigations of complexes in solution and for studies in non-aqueous solvents.

Polar Solvent. A solvent, the molecules of which have good solvating power. Generally, the greater the dipole of the molecules of the solvent the better the liquid as a solvent for ionic substances.

Polaroid. A doubly refracting material that plane-polarizes unpolarized light passed through it. It consists of a plastic sheet strained in a manner that makes it birefringent by aligning its molecules. Sunglasses incorporating a Polaroid material absorb light that is vibrating horizontally—produced by reflection from horizontal surfaces—and thus reduce glare.

Polydispersion. In most colloidal sols the dispersed particle sizes vary within very wide limits unless special means are adopted to exclude particles outside certain size limits. Such sols are said to be polydispersed. Gelatin solution is a poly-disperse protein dispersion with particles of mol. wt. between 10000 and 70000.

Polyelectrolyte. A macromolecular compound containing many ionizable groups within the same molecule. A polyelectrolyte may be purely anionic or cationic, or may be amphoteric; the individual groups may be weak or strong acids. *E.g.*, strongly acidic—polystyrene sulphonic acid; weakly acidic—polymethacrylic acid; strongly basic—polyvinyl-pyridinium bromide; weakly basic—polyvinyl-amine; amphoreric—polyglycine, proteins in general. Although the term is usually used for soluble substances, ion-exchange resins and fibrous proteins could be considered insoluble polyelectrolytes.

Polymorphism. The existence of a substance in more than one crystal form, *e.g.*, HgI_2 yellow-orthorhombic, red-tetragonal.

The different forms are referred to as polymorphs. Substances which exist in two forms are referred to as dimorphic. Polymorphism of the elements is known as allotropy. Thermodynamically unstable forms (*e.g.*, the yellow form of HgI_2) are referred to as metastable.

Positive Ray Analysis. When a beam of positive particles (rays) is passed through a magnetic and an electric field, the various ions are deflected to varying extents depending upon their velocities, masses and charges. Such analysis is the basis of a mass spectrometer.

Positive Rays. Also called canal rays. When a gas at low pressure is subjected to an electric discharge some of the molecules lose one or more electrons and form positive ions. These ions pass towards the cathode under the influence of the electric field within the discharge tube. If a hole is cut in the cathode they will pass through it in the form of a positive ion beam, which is called a positive ray.

Potential Barrier. A region containing a maximum of potential that prevents a particle on one side of it from passing to the other side. According to classical theory a particle must possess energy in excess of the height of the potential barrier to pass it. However, in quantum theory there is a finite probability that a particle with less energy will pass through the barrier (see tunnel effect. A potential barrier surrounds the atomic nucleus and is important in nuclear physics ; a similar but much lower barrier exists at the interface between semiconductors and metals and between differently doped semiconductors. These barriers are important in the design of electronic devices.

Potentiometric Titration. At the end-point in a titration there is a rapid change in the concentration of all of the reacting species. An inert electrode immersed in the solution thus shows a rapid change in e.m.f. as the end-point is approached and the end-point can be determined by observing the change in e.m.f. of the electrode during the titration. This method is especially useful when coloured solutions are used and the use of indicators is impractical.

Poundal. Symbol : pdl. The unit of force in the f.p.s. system. It is equal to 0.138255 N.

Predissociation. A term used in spectroscopy in connection with a certain type of molecular band spectrum, in which the bands are diffuse, showing no sharp convergence limit but merge gradually into a continuum. The behaviour is explained by supposing that the absorption of light excites the molecule initially to a metastable state, but after a few vibrations the molecule may dissociate.

Pressure. Symbol : p. The pressure on a surface due to forces from another surface or from a fluid is the force acting at 90° to unit area of the surface :

$$\text{pressure} = \text{force}/\text{area}$$

The unit is the pascal (Pa).

Pressure Gauge. Any device used to measure pressure. Three basic types are in use : the liquid-column gauge (*e.g.*, the mercury barometer and the manometer), the expanding-element gauge (*e.g.*, the Bourdon gauge and the aneroid barometer), and the electrical transducer. In the last category the strain gauge is an example. Capacitor pressure gauges also come into this category. In these devices, the pressure to be measured displaces one plate of a capacitor and thus alters its capacitance.

Primary Cell. A voltaic cell in which the chemical reaction that produces the e.m.f. is not reversible. *Compare* accumulator.

Primitive Lattice. A lattice having only one equivalent point in the unit cell. Often a primitive lattice is expressed with the atom of molecule corresponding to the point at the corners of the unit cell.

Promoter (Activator). A substance that improves the efficiency of a catalyst. It does not itself catalyse the reaction but assists the catalytic activity. For example, alumina or molybdenum promotes the catalytic activity of finely divided iron in the Haber process.

The manner in which a promoter functions is not fully understood ; no one theory covers all the examples.

See coenzymes.

Protective Colloids. Hydrophilic colloids *e.g.*, gelatin, being themselves unaffected by small concentrations of electrolytes, due to hydration are able, when added in very small amounts to protect hydrophobic sols from the coagulating influence of electrolytes. These are therefore termed protective colloids.

All hydrophilic colloids possess some degree of protective action and gelatin, starch and casein are used commercially for this purpose.

Proton. An elementary particle that is stable, bears a positive charge equal in magnitude to that of the electron, and has a mass of 1.672614×10^{-27} kg, which is 1836.12 times that of the electron. The proton is a hydrogen ion and occurs in all atomic nuclei.

Protonation. Addition of a proton, *e.g.*, H_2O is protonated by HCl to form $[H_3O]^+Cl^-$.

Protonic Acid. Molecules which give H^+ by dissociation, *e.g.*, HCl, H_2SO_4.

Proton Number. See atomic number.

Pseudo First Order. Describing a reaction that appears to exhibit first-order kinetics under special conditions, even though the 'true order is greater than one. For example, in the hydrolysis of an ester in the presence of a *large* volume of water, the concentration of water remains approximately constant. The rate of reaction is thus found experimentally to be proportional to the concentration of the ester only (even though it also depends on the amount of water present). Such a reaction is described as 'bimolecular of the first order'.

Purple of Cascius. See gold.

Pyroelectricity. The property of certain crystals, such as tourmaline, of acquiring opposite electrical charges on opposite faces when heated. In tourmaline a rise in temperature of 1 K at room temperature produces a polarization of some 10^{-5} cm^{-2}.

Pyrometer. An instrument used in the chemical industry to measure high temperature, *e.g.*, in reactor vessels.

Pyrometry. The measurement of high temperatures (beyond the range of thermometers) using a *pyrometer*. Modern *narrow-band* or *spectral* pyrometers use infrared-sensitive photoelectric cells behind filters that exclude visible light. In the *optical pyrometer* (or disappearing filament pyrometer) the image of the incandescent source is focused in the plane of a tungsten filament that is heated electrically.

Pyrosols. The electrolysis of fused salts results in cloudy solutions called pyrosols which are generally regarded as colloidal, the free metal being the dispersed material. Pyrosols are easily produced by dissolving the material directly in the fused salt, *e.g.*, zinc in fused zinc chloride. Coloured glasses are sometimes formed by cooling pyrosols.

Q

Quadrupele Momont. The coupling of the nuclear quadrupole with the asymmetric electron fields in compounds. Measurement of quadrupole moments yields information about the electronic distribution over the atoms within a molecule or ion.

Quadruple Point. The unique conditions of temperature and pressure at which four phases of a two-component system exist in equilibrium.

Qualitative Analysis. Analysis for the identification of constituents.

Quantitative Analysis. Analysis for the estimation of constituents.

Quantum (plural *quanta*). A definite amount of energy released or absorbed in a process. Energy often behaves as if it were 'quantized' in this way. The quantum of electromagnetic radiation is the photon.

For radiation of frequency v, the unit of energy, the quantum, is equal to hv, where h is Planck's constant. When a quantum has particle properties it is known as a photon.

Quantum Efficiency. In radiation induced processes the actual number of species which are decomposed or reacted per quantum of energy absorbed.

Quantum Electrodynamics. The use of quantum mechanics to describe how particles and electromagnetic radiation interact.

Quantum Mechanics. See quantum theory.

Quantum Theory. A mathematical theory originally introduced by Max Planck (1900) to explain the radiation emitted from hot bodies. Quantum theory is based on the idea that energy (or certain other physical quantities) can be changed only in certain discrete amounts for a given system. Other early applications were the explanations of the photoelectric effect and the Bohr theory of the atom.

Quantum Mechanics. As a system of mechanics that developed from quantum theory and is used to explain the behaviour of atoms, molecules, etc. In one form it is based on de Broglie's idea that particles can have wavelike properties—this branch of quantum mechanics is called *wave mechanics*. Wave mechanics is the basis of modern theory. See orbital.

Quantum States. States of an atom, electron, particle, etc., specified by a unique set of quantum numbers. For example, the hydrogen atom in its ground state has an electron in the K shell specified by the four quantum numbers: $n=1$, $l=0$, $m=0$, $m_s=1/2$. In the helium atom there are two electrons :

$$n=1,\ l=0,\ m=0,\ m_s=1/2$$
$$n=1,\ l=0,\ m=0,\ m_s=-1/2.$$

Quinhydrone Electrode. A half-cell consisting of a platinum electrode in an equimolar solution of quinone (cyclohexadiene-1,4-dione) and hydroquinone (benzene-1,4-diol). It depends on the oxidation-reduction reaction

$$C_6H_4(OH)_2 \rightleftharpoons C_6H_4O_2+2H^++2e.$$

Quantum Statistics. A statistical description of a system of particles that obeys the rules of quantum mechanics rather than classical mechanics. In quantum statistics, energy states are considered to be quantized ; if the particles are treated as indistinguishable. *Bose-Einstein statistics* apply and the particles are called *bosons*. All known bosons have an angular momentum nh, where n is zero or an integer and h is the Planck constant. For identical bosons the wave function is always symmetric. For the particular case in which not more than one particle may appear in any of the cells into which the particles are distributed, *Fermi-Dirac statistics* apply and the particles are called *fermions*. All known fermions have a total angular momentum $(n+\frac{1}{2})\ h'$ and any wave function that involves identical fermions is always antisymmetric.

R

Racah Parameters. The parameters used to express quantitatively the inter-electronic repulsion between the various energy levels of an atom. Generally, expressed as B and C. The ratios between B in a compound and B in the free ion gives a measure of the nephelauxatic effect.

Rad. A unit of absorbed dose of ionizing radiation, defined as being equivalent to an absorption of 10^{-2} joule of energy in one kilogram of material.

Radian. Symbol : rad. The SI unit of plane angle ; 2π radian is one complete revolution (360°).

Radiation. In general the emission of energy from a source, either as waves (light, sound, etc.) or as moving particles (beta rays or alpha rays).

Radioactivity. The disintegration of certain unstable nuclides with emission of radiation.

Radiocarbon Dating. Carbon in living things contains a uniform amount of radioactive ^{14}C, which is produced constantly in the atmosphere by the action of cosmic radiation. Once the host dies, the exchange of nonradioactive carbon with the atmospheric ^{14}C ceases. Thus from the amount of ^{14}C in the dead sample the age of it may be determined. ^{14}C has a half-life of 5570 years and the method is useful for samples having ages up to 30 000 years.

Radiochemistry. The branch of chemistry concerned with radioactive compounds and with ionization. It includes the study of compounds of radioactive elements and the preparation and use of compounds containing radioactive atoms. See labelling ; radiolysis.

Radioisotope. A radioactive isotope of an element. Tritium, for instance, is a radioisotope of hydrogen. Radioisotopes are extensively used in research as sources of radiation and as tracers in studies of chemical reactions. Thus, if an atom in a compound is replaced by a radioactive nuclide of the element (a *label*) it is possible to follow the course of the chemical reaction. Radioisotopes are also used in medicine for diagnosis and treatment.

Radiometric Dating (Radioactive Dating). See dating techniques.

Radio Waves. A form of electromagnetic radiation with wavelengths greater than a few millimetres. See also electromagnetic radiation.

Raman Effect. When light of frequency v_0 is scattered by molecules of a substance, which have a vibrational frequency of v_i, the scattered light, when analysed spectroscopically has lines of frequency v, where

$$v = v_0 \pm v_i$$

This spectrum is called a Raman spectrum and corresponds to the vibrational or rotational changes in the molecule. The selection rules for Raman activity are different from those for i.r. activity and the two types of spectroscopy are complementary in the study of molecular structure. Modern Raman spectrometers use lasers for excitation. In the resonance Raman effect excitation at a frequency corresponding to electronic absorption causes great enhancement of the Raman spectrum.

Ramsay-Shields Equation. An equation relating the molecular surface energy of a liquid with its temperature

$$\gamma\left[\frac{M}{D}\right]^{2/3}=k(Tc-T-6)$$

where γ is the surface tension of liquid, M its molecular weight, D its density, T_c is the critical temperature and T the temperature at which the measurement is made, k is known as the Ramsay-Shields constant and for many undissociated liquids, has an approximate value of 2.12.

Raoult's Law. When a solute is dissolved in a solvent, the vapour pressure of the latter is lowered proportionally to the mole fraction of solute present. Since the lowering of vapour pressure causes an elevation of the boiling point and a depression of the freezing point, Raoult's law also applies and leads to the conclusion that the elevation of boiling point or depression of freezing point is proportional to the weight of the solute and inversely proportional to its molecular weight. Raoult's law is strictly only applicable to ideal solutions since it assumes that there is no chemical interaction between the solute and solvent molecules.

Rarefaction. A reduction in the pressure of a fluid and therefore of its density.

Rate Constant (Velocity Constant) Symbol : k. The constant in an expression for the rate of a chemical reaction in terms of concentrations (or activities). For instance, in a simple unimolecular reaction

$$A \rightarrow B$$

the rate is proportional to the concentration of A, *i.e.*

$$\text{rate}=k[A]$$

where k is the rate constant, which depends on the temperature. The equation is the *rate equation* of the reaction, and its form depends on the reaction mechanism.

Rate Determining Step (limiting step). The slowest step in a multistep reaction. Many chemical reactions are made up of a number of steps in which the one with the lowest velocity is the one that determines the rate of the overall process.

The overall rate of a reaction cannot exceed the rate of the slowest step. For example, the first step in the reaction between acidified potassium iodide solution and hydrogen peroxide is the rate determining step :

$$H_2O_2+I \rightarrow H_2O+OI \text{ (slow)}$$
$$H^+ + OI \rightarrow HOI \text{ (fast)}$$
$$HOI+H^+ + I \rightarrow I_2+H_2O \text{ (fast)}$$

Rate of Reaction. A measure of the amount of reactant consumed in a chemical reaction in unit time. It is thus a measure of the number of effective collisions between reactant molecules. The rate at which a reaction proceeds can be measured by the rate the reactants disappear or by the rate at which the products are formed. The principal factors affecting the rate of reaction are temperature, pressure, concentration of reactants, light, and the action of a catalyst. The units usually used to measure the rate of a reaction are mol $dm^{-3}s^{-1}$. See also law of mass action.

Rationality of Indices, Law of. See rationality of intercepts, law of.

Rationality of Intercepts, Law of The intercepts of the faces of a crystal upon the axes of the crystal bear a simple ratio to each other. This is sometimes referred to as the law of rationality of indices.

Rationalized Units. A system of units in which the equations have a logical form related to the shape of the system. SI units form a rationalized system of units. In it formulae concerned with circular symmetry contain a factor of 2π ; those concerned with radial symmetry contain a factor of 4π.

Ratio of Specific Heats. The ratio between the specific heat of a gas at constant pressure (C_p) and that at constant volume (C_v). This ratio (C_p/C_v) is characteristic of the number of atoms in each molecule of the gas ; monatomic gases have a value of 1.67 ; diatomic, 1.40 ; and triatomic 1.33.

Reactant. A compound taking part in a chemical reaction.

Reaction. See chemical reaction.

Reaction Velocity. The rate of a chemical reaction expressed either as the rate of disappearance of reactant molecules or the rate of formation of product molecules.

Real Gas. A gas that does not have the properties assigned to an ideal gas. Its molecules have a finite size and there are forces between them (see equation of state).

Redox. See oxidation-reduction.

Redox Catalyst. A combination of a free radical catalyst with a reducing ion or salt which increases the rate of free radical production. Used in polymerization reactions, which in consequence can either be carried out at lower temperatures or at faster rates. Many synthetic rubbers are produced by this method.

Redox Indicator. A substance which undergoes a definite colour change during its reversible oxidation or reduction. The substance has a definite, and preferably narrow, potential range over which the colour change occurs. *E.g.*, methylene blue is colourless when reduced, blue when oxidized.

Reduced Mass. For any system of particles of masses m and M in harmonic motion about each other, the reduced mass μ is given by

$$\mu = \frac{mM}{(m+M)}.$$

Redwood Viscometer. An instrument widely used in the petroleum industry for determination of viscosity. The viscosity is measured in Redwood seconds, the time taken for 50 cm^3 of oil to flow through a standard orifice.

Recently the tendency has been for Redwood viscosity measurements to be replaced by kinematic viscosities.

Reference Electrode. A standard electrode used in electrochemistry. Examples are the calomel electrode, Ag—AgCl, $Hg-Hg_2SO_4$ electrodes.

Relative Atomic Mass. Symbol : A_r. The ratio of the average mass per atom of the naturally occurring element to 1/12 of the mass of an atom of nuclide ^{12}C. It was formerly called *atomic weight.*

Relative Density (R.D.). The ratio of the density of a substance to the density of some reference substance. For liquids or solids it is the ratio of the density (usually at 20°C) to the density of water (at its maximum density). This quantity was formerly called *specific gravity.* Sometimes relative densities of gases are used ; for example, relative to dry, both gases being at s.t.p.

Relative Molecular Mass (Molecular Weight). Symbol M_r. The ratio of the average mass per molecule of the naturally occurring form of an element or compound to 1/12 of the mass of a carbon-12 atom. It is equal to the sum of the relative atomic masses of all the atoms that comprise a molecule.

Relaxation. The process of loss of energy from an excited state to another excited state or to the ground state.

Rem. See radiation units.

Resonance. In the optical sense the term refers to the absorption of radiation (visible, i.r., micro-wave) by a system which it is capable of emitting. *E.g.*, mercury vapour, when suitably excited in an electrical discharge, will emit light of wavelength λ 2536 A°. This same wavelength is very readily absorbed by Hg vapour, and the energy so absorbed may be utilized in producing photochemical reactions. The process is analogous to acoustic resonance.

The term resonance has also been applied in valency. The general idea of resonance in this sense is that if the valency electrons in a molecule are capable of several alternative arrangements which differ by only a small amount in energy and have no geometrical differences, then the actual arrangement will be a hybrid of these various alternatives. See mesomerism. The stabilization of such a system over the non-resonating forms is the resonance energy.

Resonance Ionization Spectroscopy. A technique allowing the selection, isolation and observation of as small an amount as a single distinct atom by using a laser to ionize that atom selectively.

Resonance Raman Spectroscopy. See Raman spectroscopy.

Reverse Osmosis. Separation of solute from a solution by causing the solvent to flow through a membrane at pressures higher than the normal osmotic pressure. Normally, solute and solvent molecules are about the same size : where solute molecules are significantly larger than the solvent molecules the term ultrafiltration is used. The most commonly used membrane is cellulose acetate although new, improved materials have been suggested for specific purposes.

Reverse osmosis is used for desalination of seawater, treatment of recycle water in chemical plants and separation of industrial wastes. More recently the technique has been applied to concentration and dehydrogenation of food products such as milk and fruit juices.

Reversible Change. A change in the pressure, volume, or other properties of a system, in which the system remains at equilibrium throughout the change. Such processes could be reversed ; *i.e.*, returned to the original starting position through the same series of stages. They are never realized in practice. An isothermal reversible compression of a gas, for example, would have to be carried out infinitely slowly and involve no friction, etc. Ideal energy transfer would have to take place between the gas and the surroundings to maintain a constant temperature.

In practice, all real processes are ***irreversible changes*** in which there is not an equilibrium throughout the change. In an irreversible change, the system can still be returned to its original state, but not through the same series of stages. For a closed system, there is an entropy increase involved in an irreversible change.

Reversible Process. Any process in which the variables that define the state of the system can be made to change in such a way that they pass through the same values in the reverse order when the process is reversed. It is also a condition of a reversible process that any exchanges of energy, work, or matter with the surroundings should be reversed in direction and order when the process is reversed. Any process that does not comply with these conditions when it is reversed is said to be an *irreversible process.* All natural processes are irreversible, although some processes can be made to approach closely to a reversible process.

Rhe. A unit of fluidity equal to the reciprocal of the poise.

Rheolcgy. The study of the defcrmation and flow of matter.

Rheopexy. The phenomenon whereby the gelation of some thixotropic sols is accelerated by gentle mechanical agitation. An example is gypsum-water paste which, if allowed to stand, solidifies in about 10 minutes, but when gently agitated solidifies in seconds. Vanadium pentoxide sol exhibits similar behaviour.

Roentgen. Symbol : R. A unit of radiation, used for X-rays and γ-rays, defined in terms of the ionizing effect on air. One roentgen induces 2.58×10^{-4} coulomb of charge in one kilogram of dry air.

Rontgen Rays. An alternative name for X-rays, which were discovered by Rontgen in 1895.

Rotatory Power, Specific. The specific rotatory power of a pure liquid is given by

$$[a]^t\text{D} = (a/ld)$$

where $[a]^t\text{D}$ is the specific rotatory power at temperature t for the sodium D-line ; a is the rotation of the plane of polarization produced by a column of liquid of length l (decimetres) and density d. For solutions the corresponding equation is

$$[a]^t\text{D} = 100a/lc$$

where c is the solute concentration (g/ml).

The molecular rotation is given by

$$[\text{M}]^t\text{D} = \text{M}[a]/100$$

where M is the molecular weight.

Russell-Saunders Coupling, LS Coupling. A system of arriving at the different energy levels associated with unfilled shells, in which electron spins are assumed to couple only with electron spins, and orbital momenta to couple only with orbital momenta. Used to describe the ions of the first row of the transition elements. The heavier elements require the use of jj coupling in which spin-orbital interaction is assumed.

Rusting. Corrosion of iron (or steel) to form a hydrated iron (III) oxide $Fe_2O_3 \cdot {}_xH_2O$. Rusting occurs only in the presence of both water and oxygen. It is an electrochemical process in which different parts of the iron surface act as electrodes in a cell reaction. At the anode, iron atoms dissolve as Fe^{2+} ions :

$$Fe(s) \rightarrow Fe^{2+}(aq) + 2e$$

At the cathode, hydroxide ions are formed :

$$O_2(aq) + 2H_2O(l) + 4e \rightarrow 4OH^-(aq)$$

The $Fe(OH)_2$ in solution is oxidized to Fe_2O_3. Rusting is accelerated by impurities in the iron and by the presence of acids or other electrolytes in the water.

S

Sacrificial Protection (Cathodic Protection). The protection of iron or steel against corrosion (see rusting) by using a more reactive metal. A common form is galvanizing (see galvanized iron), in which the iron surface is coated with a layer of zinc. Even if the zinc layer is scratched, the iron does not rust because zinc ions are formed in solution in preference to iron ions. Pieces of magnesium alloy are similarly used in protecting pipelines, etc.

Salt. A compound formed by reaction of an acid with a base, in which the hydrogen of the acid has been replaced by metal or other positive ions. Typically, salts are crystalline ionic compounds such as Na^+Cl^- and $NH_4^+NO_3^-$. Covalent metal compounds, such as $TiCl_4$, are also often regarded as salts.

Salt Bridge. An electrical connection made between two half cells. It usually consists of a glass U-tube filled with agar jelly containing a salt, such as potassium chloride. A strip of filter paper soaked in the salt solution can also be used.

Salting Out. A term used for both colloidal and non-colloidal solutions. In the context of colloids, salting out refers to the coagulation of hydrophilic colloids, *e.g.* gelatin or soaps, by the addition of large concentrations of electrolytes. This should not be confused with the coagulation of hydrophobic sols by small quantities of electrolytes. With non-colloidal solutions the term salting out is used to describe the reduction in solubility of a non-electrolyte by addition of a strong electrolyte.

Saturated Solution. Refers to a solution that contains the maximum equilibrium amounts of solute at a given temperature. A solution is saturated if it is in equilibrium which its solute. If a saturated solution of a solid is cooled slowly, the solid may stay temporarily in solution ; *i.e.*, the solution may contain *more* than the equilibrium amount of solute. Such solutions are said to be *supersaturated.*

Saturated Vapour. Refers to a vapour that is in equilibrium with the solid or liquid. A saturated vapour is at the maximum pressure (the saturated vapour pressure) at a given temperature. If the temperature of a saturated vapour is lowered, the vapour condenses. Under certain circumstances, the substance may stay temporarily in the vapour phase ; *i.e.* the vapour contains *more* than the equilibrium concentration of the substance. The vapour is then said to be *supersaturated.*

Saturation. See supersaturation.

Scintillation Counting. The detection and quantitative estimation of radioactivity by the scintillations produced when the ionizing radiation interacts with a scintillating medium. This may be a screen of material such as zinc sulphide (for α-particles), a lithium-doped sodium iodide crystal (for γ-rays) or liquid scintillators (β-particles).

Second Symbol : s. The SI base unit of time. It is defined as the duration of 9,192.631,770 cycles of a particular wavelength of radiation corresponding to a transition between two hyperfine levels in the ground state of the caesium-133 atom.

Secondary Cell. Refers to a voltaic cell in which the chemical reaction producing the e.m.f. is reversible and the cell can therefore be charged by passing a current through it. See accumulator. *Compare* primary cell.

Second-order Reaction. Refers to a reaction in which the rate of reaction is proportional to the product of the concentrations of two of the reactants or to the square of the concentration of one of the reactants ; *i.e.*,

$$\text{rate}=k[\text{A}][\text{B}]$$

or

$$\text{rate}=k[\text{A}]^2$$

For example, the hydrolysis by dilute alkali of an ester is a second-order reaction :

$$\text{rate}=k[\text{ester}][\text{alkali}]$$

The rate constant for a second-order reaction has the units $mol^{-1}\ dm^3\ s^{-1}$. Unlike a first-order reaction, the time for a definite fraction of the reactants to be consumed is dependent on the original concentrations.

Secondary Radiaion. Radiation produced by the absorption of some other, generally more energetic, radiation. Thus, when X-rays strike a body it emits other X-rays, which are termed secondary rays, or secondary radiations, characteristic of the atom which acts as the emitter (see X-ray analysis). This phenomenon is also observed with electrons and is then termed secondary emission.

Sedimentation Gravitational settling of solid particles suspended in a liquid so that the original suspension is divided into a reasonably clear effluent and a sludge with an increased solid content. See clarification, thickening.

As the particles of a suspension obey Stokes's law their rate of sedimentation through a liquid of known viscosity gives a measure of their size. Although colloidal particles do not settle appreciably under the influence of gravity, they can be made to do so in a high-speed centrifuge.

Sedimentation Potential. The measurable potential difference which results from the motion of the disperse phase through the dispersion medium, when the particles of a supension are allowed to settle under the influence of gravity.

Seger Cones (Pyrometric Cones). Refer to a series of cones which are used to indicate the temperature inside a furnace or kiln. The cones are made from different mixtures of clay, limestone, feldspars, etc., and each one softens at a different temperature. The drooping of the vertex is an indication that the known softening temperature has been reached and thus the furnace temperature can be estimated.

Selection Rules. Rules derived from group theory which denote, *e g.* whether a particular vibration is i.r. or Raman active or a particular transition between electronic energy levels is allowed or forbidden.

Selenium Cell. Either of two types of photoelectric cell ; one type relies on the photoconductive effect, the other on the photovoltaic effect (see photoelectric effect). In the photoconductive selenium cell an external e.m.f. must be applied ; as the selenium changes its resistance on exposure to light, the current produced is a measure of the light energy falling on the selenium. In the photovoltaic selenium cell, the e.m.f. is generated within the cell. In this type of cell, a thin film of vitreous of metallic selenium is applied to a metal surface, a transparent film of another metal, usually gold or platinum, being placed over the selenium. Both types of cell are used as light meters in photography.

Semi-conductors. Imperfections or impurities in what are normally insulators may give rise to a temperature dependent conductivity (metallic conductivity decreases with rise in temperature) arising because the highest occupied energy level is very close to an unoccupied level. Such materials are called semi-conductors. There are two main types $MX_{1-x}(e)_x$ with an electron (*e*) in an anion vacancy, *e.g.* (NaCl) and $M_{1+x}X$-$(e)_x$ with an interstitial M and an electron in its vicinity (*e.g.* ZnO) which conduct by movements of the electrons and are called *n* type. $(+)_xM_{1-x}X$ with a vacant metal site and a neighbouring +ve charge (M^{2+}) in the vicinity (*e.g.* Cu_2O) and $(+)_xMX_{1+x}$ with an interstitial anion and a neighbouring +ve charge (M^{2+}) (*e.g.* VO_2) conduct formally by movement of the +ve charge (actually by electron movement) and are called *p* type. Semi-conductors are of great importance in electronics and are used in, *e.g.* transistors, thermistors, solid rectifiers and microprocessors.

Semipermeable Membrane. A membrane that is permeable to molecules of the solvent but not the solute in osmosis. Semipermeable membranes can be made by supporting a film of material (*e.g.* cellulose) on a wire gauze or porous pot.

Series, Spectroscopic. A group of lines the frequencies of which are simply related to each other and may be expressed by a simple formula (*e.g.* the Balmer series).

SI. SI units.

Side Reactions. Few reactions proceed quantitatively to one product. Generally a number of reactions occur. The predominant reaction is the main reaction ; other reactions are termed side reactions.

Siemens (Mho) Symbol. S. The SI unit of electrical conductance, equal to a conductance of one ohm^{-1}.

Sievert. The SI unit of dose equivalent (see radiation units).

SI Units. (Systeme International d'Units). The internationally adopted system of units used for scientific purposes. It has seven base units (the metre, kilogram, second, kelvin, ampere, mole, and candela) and two supplementary units (the radian and steradian). Derived units are formed by multiplication and/or division of base units ; a number have special names. Standard prefixes are used for multiples and submultiples of SI units. The SI system is a coherent rationalized system of units.

Base and Supplementary SI Units

Physical quantity	*name of SI unit*	*symbol for unit*
length	metre	m
mass	kilogram(me)	kg
time	second	s
electric current	ampere	A
thermodynamic temperature	kelvin	K
luminous intensity	candela	cd
amount of substance	mole	mol
plane angle	radian	rad
solid angle	steradian	sr
supplementary units		

Derived SI Units with Special Names

Physical quantity	*name of SI unit*	*symbol for SI unit*
frequency	hertz	Hz
energy	joule	J
force	newton	N
power	watt	W
pressure	pascal	Pa
electric charge	coulomb	C
electric potential difference	volt	V
electric resistance	ohm	Ω
electric conductance	siemens	S
electric capacitance	farad	F
magnetic flux	weber	Wb
inductance	henry	H
magnetic flux density	tesla	T
luminous flux	lumen	lm
illuminance (illumination)	lux	lx
absorbed dose	gray	Gy

Decimal Multiples and Submultiples to be used with SI Units

Submultiple	*prefix*	*symbol*	*multiple*	*prefix*	*symbol*
10^{-1}	deci-	d	10^{1}	deca-	da
10^{-2}	centi-	c	10^{2}	hecto-	h
10^{-3}	milli-	m	10^{3}	kilo-	k
10^{-6}	micro-	μ	10^{6}	mega-	M
10^{-9}	nano-	n	10^{9}	giga-	G
10^{-12}	pico-	p	10^{12}	tera-	T
10^{-15}	femto-	f	10^{15}	peta-	P
10^{-18}	atto-	a	10^{18}	exa-	E

Slip Planes. The planes of weakness of crystals corresponding to crystalline boundaries and to planes of atoms with only weak forces between such planes (*e.g.* the layers in graphite).

Smectic. See liquid crystal.

Smoke. A fine suspension of solid particles in a gas.

Sol. A colloid consisting of solid particles distributed in a liquid medium. A wide variety of sols are known ; the colours often depend markedly on the particle size. The term *aerosol* is used for solid or liquid phases dispersed in a gaseous medium. See *also* colloid.

Solid. Refers to a state of matter in which there is a three-dimensional regularity of structure, resulting from the proximity of the component atoms, ions, or molecules and the strength of the forces between them. True solids are crystalline (see *also* amorphous). If a crystalline solid is heated, the kinetic energy of the components increases. At a specific temperature, called the *melting point*, the forces between the components become unable to contain them within the crystal structure. At this temperature, the lattice breaks down and the solid becomes a liquid.

Solid Solution. A solid solution is formed when two or more elements or compounds share a common lattice. The composition may vary within very wide limits. A substitutional alloy results when atoms of solute replace atoms of solvent in its crystal lattice, *e.g.* nickel copper alloys. An interstitial solid solution occurs when a small atom, such as carbon, enters the interstices of the host lattice of larger atoms such as metals.

Solid State Electrode. Ion-selective electrodes.

Solubility. Refers to the amount of one substance that could dissolve in another to form a saturated solution under specified conditions of temperature and pressure.

Solubility is measured in kilograms per metre cubed, moles per kilogram of solvent, etc. The solubility of a substance in a given solvent depends on the temperature. Generally, for a solid in a liquid, solubility increases with temperature ; for a gas, solubility decreases.

Solubility Curve. A graphical representation of the variation of the solubility with temperature.

Solubility Product. When a saturated solution of an electrolyte is in contact with some undissolved electrolyte the following equilibria may be considered

$$AB_{solid} \rightleftharpoons AB_{dissolved} \rightleftharpoons A^+ + B^-$$

the equilibrium constant for this process may be written as

$$K = \frac{[A^+][B^-]}{[AB]_{dissolved}}$$

However, because of the equilibrium between the solid and dissolved undissociated AB, the concentration of the latter may be considered to remain constant and hence

$$[A^+][B^-] = K_{sp}$$

where K_{sp}=solubility product of the electrolyte, It can be seen that if an excess of A^+ of B^- ions are added to the solution, the K_{sp} is exceeded and the salt will be precipitated to restore the equilibrium.

Solubility products are meaningful only for sparingly soluble salts. If the product of ions exceeds the solubility product, precipitation occurs.

Solute. The substance dissolved in a solvent in forming a solution.

Solution. A liquid system of two or more species that are intimately dispersed within each other at a molecular level. The system is therefore totally homogeneous. The major component is called the solvent (generally liquid in the pure state) and the minor component is called the solute (gas, liquid, or solid).

The process occurs because of a direct intermolecular interaction of the solvent with the ions or molecules of the solute. This interaction is called solvation. Part of the energy of this interaction appears as a change in temperature on dissolution. See *also* solid solution, solubility.

Solvation. The interaction of ions of a solute with the molecules of solvent. For instance, when sodium chloride is dissolved in water the sodium ions attract polar water molecules, with the negative oxygen atoms pointing towards the positive Na^+ ion. Solvation of transition-metal ions can also occur by formation of coordinate bonds, as in the hexaquocopper(II) ion $[Cu(H_2O)_6]^{2+}$. Solvation is the process that causes ionic solids to dissolve, because the energy released compensates for the energy necessary to break down the crystal lattice. It occurs only with polar solvents. Solvation in which the solvent in water is called *hydration*.

Solvent. A liquid capable of dissolving other materials (solids, liquids, or gases) to form a solution. The solvent is generally the major component of the solution.

Solvent Extraction. The process of separating one constituent from a mixture by dissolving it in a solvent in which it is soluble but in which the other constituents of the mixture are not. The process is usually carried out in the liquid phase, in which case it is also known as *liquid-liquid extraction*. In liquid-liquid extraction, the solution containing the desired constituent must be immiscible with the rest of the mixture. The process is widely used in extracting oil from oil-bearing materials.

Solvolysis. A reaction between a compound and its solvent. See hydrolysis.

Sorption. Absorption of gases by solids.

Sorption Pump. A type of vacuum pump in which gas is removed from a system by absorption on a solid (*e.g.* activated charcoal or a zeolite) at low temperature.

Space Group. The whole array of symmetry-elements in a crystal framework or lattice. The total number of possible space groups is 230. The atoms in any crystal must be arranged so that they fit into one of these groups. The framework composing a space group is termed a *space lattice*.

Space Lattice. A term applied to the array of points in any crystalline structure for which the pattern repeats. By joining these points the crystals can be divided into a series of parallel *unit cells*, each of which contains a complete unit of pattern. The atoms composing the crystal occupy the points in the lattice so that a space lattice affords a convenient means of representing the atomic structure of a crystalline substance. The position of the points relative to each other is found by means of X-ray analysis of the crystal.

Specific. Denoting a physical quantity per unit mass. For example, volume (V) per unit mass (m) is called specific volume :

$$V = V_m$$

In certain physical quantities the term does not have this meaning : for example, *specific gravity* is more properly called relative density.

Specific Activity. See activity.

Specific Conductance. See conductivity.

Specific Refractivity. The quantity (r) defined by

$$r = \frac{n^2 - 1}{n^2 + 2} \cdot \frac{1}{d}$$

where n is the refractive index of the substance of density d. The molar refractivity is equal to the product of the specific refractivity and the molecular weight of the substance.

Specific Heat Capacity. See heat capacity.

Specific Rotatory Power G Symbol : α_m. The rotation of plane-polarized light in degrees produced by a 10 cm length of solution containing 1 g of a given substance per millilitre of stated solvent. The specific rotatory power is a measure of the optical activity or substances in solution. It is measured at 20°C using the D-line of sodium.

Spectrograph. See spectroscope.

Spectrophotometer. See spectrometer.

Spectroscope. An instrument for examining the different wavelengths present in electromagnetic radiation. See *also* spectrometer.

Spectroscopy. 1. The production and analysis of spectra. There are many spectroscopic techniques designed for investigating the electromagnetic radiation emitted or absorbed by substances. Spectroscopy, in various forms, is used for analysis of mixtures, for identifying and determining the structures of chemical compounds, and for investigating energy levels in atoms, ions, and molecules. In the visible and longer wavelength ultraviolet, transitions correspond to electronic energy levels in atoms and molecules. The shorter wavelength ultraviolet corresponds to transitions in ions. In the X-ray region, transitions in the inner shells of atoms or ions are involved. The infrared region corresponds to vibrational changes in molecules, with rotational changes at longer wavelengths.

2. Any of various techniques for analysing the energy spectra of beams of particles or for determining mass spectra.

Spectrum. 1. Refers to a distribution of entities or properties arrayed in order of increasing or decreasing magnitude. For example, a beam of ions passed through a mass spectrograph, in which they are deflected according to their charge-to-mass ratios, will have a range of masses called a *mass spectrum*. A *sound spectrum* is the distribution of energy over a range of frequencies of a particular source. 2. Refers to a range of electromagnetic energies arrayed in order of increasing or decreasing wavelength or frequency (see electromagnetic spectrum). The *emission spectrum* of a body or substance is the characteristic range of radiations it emits when it is heated, bombarded by electron or ions, or absorbs photons. The *absorption spectrum* of a substance may be produced by examining, through the substance and through a spectroscope, a continuous spectrum of radiation. The energies removed from the continuous spectrum by the absorbing medium show up as black lines or bands; with a substance capable of emitting a spectrum these are in exactly the same positions in the spectrum as the emission lines and bands would occur in the emission spectrum.

Emission and absorption spectra may show a *continuous spectrum*, a *line spectrum*, or a *band spectrum*. A continuous spectrum contains an unbroken sequence of frequencies over a relatively wide range; it is produced by incandescent solids, liquids, and compressed gases. Line spectra are discontinuous lines produced by excited atoms and ions as they fall back to a lower energy level. Band spectra (closely grouped bands of lines) are characteristic of molecular gases or chemical compounds. See *also* spectroscopy.

Sputtering. Refers to the process by which some of the atoms of an electrode (usually a cathode) are ejected as a result of bombardment by heavy positive ions. Although the process is generally unwanted, it can be used to produce a clean surface or to deposit a uniform film of a metal on an object in an evacuated enclosure.

Stabilizer. A substance which is added to prevent chemical change (*i.e.*, a negative catalyst).

Stalagmometer. Apparatus for comparing the surface tensions or interfacial tensions of liquids by comparing the masses of drops formed at a given orifice.

Standard Cell. A voltaic cell, such as a Clarke cell, or Weston cell, which is used as a standard of e.m.f.

Standard Electrode. An electrode which produces a standard e.m.f. Electrode potentials are normally quoted relative to the hydrogen electrode, although in practice reference is usually made to the calomel electrode.

Standard Electrode Potential. See electrode potential.

Standard Pressure. An internationally agreed value ; a barometric height of 760 mmHg at 0°C : 101325 Pa (approximately 100 kPa).

This is sometimes called the *atmosphere* (used as a unit of pressure). The *bar*, used mainly when discussing the weather, is 100 kPa exactly.

Standard Rate. Using thermodynamics it is possible to predict whether or not reaction will occur when substances are placed together in a closed system. However, to do this it is necessary to define a standard state of the system as one which contains all the reactants and products : (1) in the pure state if they are crystalline substances or liquids ; (2) at one atmosphere pressure if they are gases ; (3) at a concentration of 1F (one formula weight in grams per litre of solution) if they are dissolved in a solvent ; and (4) at a specified temperature, usually, but not necessarily, 298 K.

Thermodynamic quantities which refer to the standard state are denoted by superscript zeros (^{0}), *e.g.*, ΔG_T^0, ΔH_T^0, ΔS_T^0, the subscript denoting the temperature T of the system.

Standard State. A state of a system used as a reference value in thermodynamic measurements. Standard states involve a reference value of pressure (usually 1 atmosphere, 101.325 kPa) or concentration (usually 1 M). Thermodynamic functions are designated as 'standard' when they refer to changes in which reactants and products are all in their standard and their normal physical state. For example, the standard molar enthalpy of formation of water at 298 K is the enthalpy change for the reaction

$$H_2(g) + \tfrac{1}{2}O_2(g) \rightarrow H_2O(l)$$

$\Delta H_{298} = -285.83$ kJ mol^{-1}. Note the superscript is used to denote standard state and the temperature should be indicated.

Standard Tamperature. An internationally agreed value for which many measurements are quoted. It is the melting temperature of water, 0°C (273.15 K).

Standard Temperature and Pressure, STP. Arbitrary reference conditions of 1 standard atmosphere at 0°C. This is represented by 1.013×10^5 N/m^2 and 273.15 K.

State of Matter. One of the three physical states in which matter can exist, *i.e.*, solid, liquid, or gas. Plasma is sometimes regarded as the fourth state of matter.

Stationary State. In a reversible chemical reaction, a condition is reached at which the rate of the forward reaction is exactly equal to the rate of the backward reaction. At this point the reaction system has reached a state of equilibrium often referred to as a stationary state In the stationary state the concentration of the reacting species remain constant.

Statistical Mechanics. The branch of physics in which statistical methods are applied to the microscopic constituents of a system in order to predict its macroscopic properties. The earliest application of this method was Boltzmann's attempt to explain the thermodynamic properties of gases on the basis of the statistical properties of large assemblies of molecules.

In classical statistical mechanics, each particle is regarded as occupying a point in *phase space*, *i.e.*, to have an exact position and momentum at any particular instant. The probability that this point will occupy any small volume of the phase space is taken to be proportional to the volume. The

Maxwell-Boltzmann law gives the most probable distribution of the particles in phase space.

With the advent of quantum theory, the exactness of these premises was distributed (by the Heisenberg uncertainty principle). In the quantum statistics that evolved as a result, the phase space is divided into cells, each having a volume h^f, where h is the Planck constant and f is the number of degrees of freedom of the particles. This new concept led to Bose-Einstein statistics, and for particles obeying the Pauli exclusion principle, to Fermi-Dirac statistics.

Steam Distillation. Refers to a method of distilling liquids that are immiscible with water by bubbling steam through them. It is dependent on the fact that the vapour pressure (and hence the boiling point) of a mixture of two immiscible liquids is lower than the vapour pressure of either pure liquid.

Steam Point. Refers to the temperature at which the maximum vapour pressure of water is equal to the standard atmospheric pressure (101325 Pa). On the Celsius scale it has the value 100°C.

Steradian. Symbol : sr. The SI unit of solid angle. The surface of a sphere, for example, subtends a solid angle of 4π at its centre. The solid angle of a cone is the area intercepted by the cone on the surface of a sphere of unit radius.

Stere. A unit of volume equal to 1 m^3. It is not now used for scientific purposes.

Stokes's Law. When a particle falls through a liquid the limiting rate of fall depends upon the difference in density of the two phases, the viscosity of the liquid, the particle size and the gravitational constant (g). The factors are related by Stokes's law :

$$v=\frac{2r^2(s-s')g}{9\eta}$$

where v is the velocity of sedimentation of the particle of radius r and density s in the liquid of density s' and viscosity η. For colloidal particles the velocity v is found to be so small as to be more than counteracted by convection, etc. Consequently a colloidal sol can never separate out without aggregation of the particles.

Stopped Flow Spectrophotometry. A technique used to investigate the rates of fast chemical reactions, with half-lives in the range 10 seconds to 5×10^{-3} seconds. The technique involves measuring the optical density of a segment of the resulting solution one or two milliseconds after the reactants have been mixed.

Storage Battery. See accumulator.

Stoichiometric. Describing chemical reactions in which the reactants combine in simple whole-number ratios.

Stoichiometric Mixture. A mixture of substances that can react to give products with no excess reactant.

S.T.P. Standard temperature and pressure.

Stream Double Refraction. Some colloidal sols such as aged vanadium pentoxide, whose dispersed particles are rod-shaped, are non-polarizing—as can be shown by placing the sol between crossed nicols, there being total extinction. When the liquid is stirred, however, or made to flow, polarization takes place, the stream lines being shown in a brightly illuminated image. This is due to the orientation of the rod-shaped molecules on stirring.

Streaming Potential. When a liquid is forced by pressure through a diaphragm a potential difference, termed the streaming potential, is set up. The streaming potential, which is the converse of electro-osmotic flow, may be measured by forcing water through a capillary of the material under investigation and determining, the potential difference between two electrodes, one placed at each end of the capillary.

Sublimation. The conversion of a solid into a vapour without the solid first melting. For instance (at standard pressure) iodine solid carbon dioxide, and ammonium chloride sublime.

At certain conditions of external pressure and temperature an equilibrium can be established between the solid phase and vapour phase.

Substrate. The substrate in an enzymic reaction is the molecule on which the enzyme acts. The three-dimensional enzyme-substrate interaction is highly favourable and necessary for the progression of the reaction.

Superconductivity. At low temperatures certain metals and a large number of intermetallic compounds possess superconductivity, have zero sensitivity and undetectable magnetic permeability. Although the origin of these properties is not fully understood, they are of considerable engineering interest.

Supercooling. The cooling of a liquid at a given pressure to a temperature below its melting temperature at that pressure without solidifying it. The liquid particles lose energy but do not spontaneously fall into the regular geometrical pattern of the solid. A supercooled liquid is in a metastable state and will usually solidify if a small crystal of the solid is introduced to act as a 'seed' for the formation of crystals. As soon as this happens, the temperature returns to the melting temperature until the substance has completely solidified.

Superheating. The raising of a liquid's temperature above its boiling temperature, by increasing the pressure.

Superlattice. See solid solution.

Supersensitization. Increasing the efficiency of spectral sensitization by introduction of a small proportion of a second dye or suitable absorbing or complexing colourless compound to an adsorbed sensitizer.

Supersaturation. 1. The state of the atmosphere in which the relative humidity is over 100%. This occurs in pure air where no condensation nuclei are available. Supersaturation is usually prevented in the atmosphere by the abundance of condensation nuclei (*e g.*, dust, sea salt, and smoke particles). 2. The state of any vapour whose pressure exceeds that at which condensation usually occurs (at the prevailing temperature).

Supplementary Units. The dimensionless units—the radian and the steradian—used along with base units to form derived units. See *also* SI units.

Surface Active Agents. Water possessing powerful intermolecular attractive forces, has a high surface tension (72.8 dynes/cm at 20°C). Many soluble substances (mainly organic) when dissolved in water reduce the surface tension even in very low concentrations.

These surface active agents have weaker intermolecular attractive forces than the solvent, and therefore tend to concentrate in the surface at the expense of the water molecules. The accumulation of adsorbed surface active agent is related to the change in surface tension according to the Gibbs adsorption equation

$$\lambda = -\frac{c}{RT}\frac{d\gamma}{dc}$$

where λ is the surface excess, c is the bulk concentration of the surface active agent and γ is the surface tension. Surface active agents, which include certain dyestuffs, wetting agents, detergents, antibiotics, bactericides, etc., may be electrically neutral, non-dissociating substances (non-ionic), or they may ionize to give surface active anions or cations with small oppositely charged counterions (usually metallic ions and halogen ions respectively).

Surface Energy. The surface energy of a system is defined as the work necessary to increase the surface by unit area against the force of surface tension.

Surface Tension. The molecules in the body of a liquid are equally attracted in all directions by the other molecules. Those at the surface, on the other hand, will have a residual inward attraction. This is associated with a *surface tension* which acts in a direction parallel to the boundary surface and tends to reduce that surface to a minimum. Work has to be done to increase the area of the surface against this force. This explains why an emulsified system always tends to be unstable and breaks down unless there is present some substance to lower the surface or interfacial tension. See surface energy.

Surface Viscosity. Unimolecular films on liquid surfaces may be either readily mobile (*e.g.*, a 'gaseous' film) or slow to flow under the action of a two-dimensional stress (see surface pressure). They therefore possess surface viscosity, the two-dimensional analogue of ordinary viscosity. It is defined and determined by analogous methods, *e.g.*, by rate of flow along a surface canal or by a rotating disc viscometer. Some surface films show surface plasticity, *i.e.*, they behave as solids until a critical shearing stress is applied.

Surfactant. A substance that lowers surface tension, *e.g.*, a soap or other detergent.

Suspension. A mixture in which small solid or liquid particles are suspended in a liquid or gas.

Swelling of Colloids. When a piece of dry gelatin is exposed to moist air or placed in water, it becomes hydrated and swells considerably. This is an effect common to all hydrophilic colloids—most natural organic matter, *e.g.*, cellulose, wood, carbohydrates. The swelling produces considerable pressure in the colloid and, in consequence, still occurs even when opposed by an external pressure of many atmospheres.

The swelling of gels is markedly affected by the presence of electrolytes, this effect being a minimum at the isoelectric point of the material. In general, sulphates, tartrates, etc. inhibit swelling, while iodides and thiocyanates promote the swelling. Thus gelatine disperses completely in iodide solution even at low temperatures.

Syneresis. The separation of liquid from a gel on standing. All gels show syneresis to some extent and the phenomenon may be regarded as a continuation of the process of gelatin, the gel material adhering at more and more places and, in consequence, shrinking and squeezing out a portion of the dispersion medium. Viscose gels synerize to a large extent. The phenomenon is commonly met in the cooking of foods and is thought to play an important part in gland secretion.

Synergism. The phenomenon by which the additive properties of two components is much greater than the sum of the effects of the two separately.

Synergist. A compound which, whilst formally inactive or weakly active, in a particular context can give enhanced activity to another compound present in the system. See, *e.g.*, super-additive developer mixtures.

Systeme International d'Unites. See SI units.

T

Tanning. The process of changing skins and hides into useful leather. Skins are cleaned and soaked in water, treated with lime, and hair and flesh removed. Neutralized with buffering salts [$(NH_4)_2SO_4$, NH_4Cl] and proteolytic enzyme and finally tanned with either tannin, obtained from vegetable sources which combine with collagen and other proteins in the hides to form leather, or with a basic chromium (III) sulphate which forms complexes with the collagen.

Tera Symbol T. A prefix which is used in the metric system to denote one milion million times. For example, 10^{12} volts = 1 teravolt (TV).

Ternary Compound. A chemical compound containing three different elements.

Tesla Symbol T. The SI unit of magnetic flux density equal to one weber of magnetic flux per square metre, *i.e.*, 1 T = 1 Wb m^{-2}. It is named after Nikola Tesla (1870-1943), Croatian-born US electrical engineer.

Thermal Capacity. See heat capacity.

Thermal Equilibrium. See equilibrium.

Thermobalance. A balance in which the sample can be heated to allow thermal analysis.

Thermochemistry. Most chemical reactions occur with the evolution or absorption of heat. Thermo-chemistry is that branch of chemistry concerned with the heat changes accompanying chemical reactions.

Thermochroism. Change in colour (and spectrum) with temperature.

Thermodynamics. The study of heat and other forms of energy and the various related changes in physical quantities such as temperature, pressure, density, etc.

The *first law of thermodynamics* states that the total energy in a closed system is conserved (constant). In all processes energy is simply converted from one form to another, or transferred from one system to another.

A mathematical statement of the first law is :

$$\delta Q = \delta U + \delta W$$

Here, δQ is the heat transferred to the system, δU the change in internal energy (resulting in a rise or fall of temperature), and δW is the external work done by the system.

The *second law of thermodynamics* can be stated in a number of ways, all of which are equivalent. One is that heat cannot pass from a cooler to a hotter body without some other process occurring. Another is the statement that heat cannot be totally converted into mechanical work, *i.e.*, a heat engine cannot be 100% efficient.

The *third law of thermodynamics* states that the entropy of a substance tends to zero as its thermodynamic temperature approaches zero.

Often a zeroth law of thermodynamics is given : that if two bodies are each in thermal equilibrium with a third body, then they are in thermal equilibrium with each other. This is considered to be more fundamental than the other laws because they assume it. See also Carnot cycle, entropy.

Thermodynamic Data. Standard enthalpies of formation ($\triangle H_f°$), standard free energies of formation ($\triangle G_f°$) and absolute entropies of selected substances at 298 K.

By definition, all pure elements have $\triangle H_f°=0$ and $\triangle G_f°=0$.

Substance	$\triangle H_f^\circ$ (*kJ mol*$^{-1}$)	$\triangle G_f^\circ$ (*kJ mol*$^{-1}$)	S° (*J mol*$^{-1}$)
$CaCO_3$ (s)	−1207	−1129	93
CH_4 (g)	−75	−51	186
C_2H_6 (g)	−85	−33	229
C_3H_8 (g)	−104	−24	270
C_6H_6 (g)	83	130	269
C_6H_6 (l)	49	125	173
C_2H_4 (g)	52	68	219
C_2H_2 (g)	227	209	201
CO (g)	−111	−137	198
CO_2 (g)	−394	−395	214
Cl_2 (g)	0	0	223
H_2 (g)	0	0	131
HCl (g)	−92	−95	187
H_2O (g)	−242	−229	189
H_2O (l)	−286	−237	70
Mg (s)	0	0	33
MgO (s)	−602	−570	27
N_2 (g)	0	0	192
O_2 (g)	0	−0	205

Thermodynamics, First Law of. This law is a consequence of the law of conservation of energy and states that mechanical energy and work are quantitatively interconvertible. Alternatively, 'In any process energy can be changed from one form to another, but it is neither created nor destroyed.'

Thermodynamics, Second Law of. This states that 'Heat cannot of itself pass from a colder to a warmer body'. Thus, if heat is to be transferred from the cold to the hot body, work must be provided by an external agency. An alternative and useful expression of the second law is : 'Any system of its own accord will always undergo change in such a way as to increase the entropy'.

Thermodynamic Temperature. Symbol : T. A temperature measured in kelvins. The size of the kelvin is the same as thet of the degree Celsium, and temperatures on the Celsius scale can be converted to thermodynamic temperatures by addine 273.15.

Thermodynamics, Third Law of. This states that 'for a perfect crystal at absolute zero on the Kelvin scale the entropy is zero'. This follows the predictions of Einstein (1907) that the specific heats of all substances would approach zero at 0 K, and the conclusions of Planck (1912) that the entropy of all pure solids and liquids approach zero at this temperature.

Thermogram. A plot of weight against temperature obtained in thermal analysis.

Thermoluminescence. Luminescence produced in a solid when its temperature is raised. It arises when free electrons and holes, trapped in a solid as a result of exposure to ionizing radiation, unite and emit photons of light. The process is made use of in *thermoluminescent dating*, which assumes that the number of electrons and holes trapped in a sample of pottery is related to the length of time that has elapsed since the pottery was fired. By comparing the luminescence produced by heating a piece of pottery of unknown age with the luminescence produced by heating similar materials of known age, a fairly accurate estimate of the age of an object can be made.

Thermoluminescent Dating. See thermoluminescence.

Thermostat. A device that controls the heating or cooling of a substance in order to maintain it at a constant temperature. It consists of a temperature sensing instrument connected to a switching device. When the temperature reaches a predetermined level the sensor switches the heating or cooling source on or off according to a predetermined program. The sensing thermometer is often a bimetallic strip that triggers a simple electrical switch. Thermostats are used for space-heating controls, in water heaters and refrigerators, and to maintain the environment of a scientific experiment at a constant temperature.

Thixotropy. The phenomenon of the isothermal gel-sol. transformation brought about by shaking or other mechanical means. It may be regarded as a packing phenomenon. *E.g.*, $Fe(OH)_3$ gels on shaking form sols which rapidly set again when allowed to stand. Thixotropy is commonly found with suspensions of clay or oil paints and emulsions.

Tonne. (**metric ton**) **Symbol** : **t.** A unit of mass equal to 10^3 kilograms (*i.e.*, one megagram).

Torr. A unit of pressure, used in high-vacuum technology, defined as 1mmHg. 1 torr is equal to 133,322 pascals. The unit is named after Evangelista Torricelli (1609-47).

Total Pressure. In a mixture of gas and vapours the total pressure exerted by the mixture is equal to the sum of the partial pressures of the constituents. See partial pressure.

Total-Radiation Pyrometer. See pyrometry.

Tracer. A substance added to a system to enable the experimenter to follow the course of a process, *e.g.*, reaction or self-diffusion, without seriously altering the conditions. More particularly, the term is chiefly used for isotopic species added as tracers, especially radioactive isotopes or rate isotopes of the lighter elements (*e.g.* ^{18}O). The former are detected and determined with the aid of a counter, the latter with a mass spectrometer. As these detectors are very sensitive, only small proportions of tracer are needed.

Transition State (**activated complex**) **Symbol.** : A short-lived high-energy molecule, radical, or ion formed during a reaction between molecules possessing the necessary activation energy. The transition state decomposes at a definite rare to yield either the reactants again or the final products. The transition state can be considered to be at the top of the energy profile.

For the reaction,

$$X+YZ=X...Y...Z\rightarrow XY+Z$$

the sequence of events is as follows. X approaches YZ and when it is close enough the electrons are rearranged producing a weakening of the bond between Y and Z. A partial bond is now formed between X and Y producing the transition state. Depending on the experimental conditions, the transition state either breaks down to form the products or reverts bank to the reactants.

Transition Temperature. The temperature at which one (allotropic) form of a substance is converted into another allotropic form. *E..g*, rhombic sulphur is converted into monoclinic sulphur at the transition temperature of 95.6°C. Below this temperature the rhombic form is the stable allotrope ; the monoclinic allotrope is the stable form above this temperature. Transition temperatures vary with the pressure exerted on the system.

Transmission Electron Microscope. See electron microscope.

Transport Number. **Symbol : t.** In an electrolyte, the transport number of an ion is the fraction of the total charge carried by that type of ion in conduction.

Triboluminescence. Luminescence caused by friction ; for example, some crystalline substances emit light when they are crushed as a result of static electric charges generated by the friction.

Triclinic. See crystal system.

Trimolecular. Describing a reaction or step that involves three molecules interacting simultaneously with the formation of a product. For example, the final step in reaction between hydrogen peroxide and acidified potassium iodide is trimolecular :

$$HOI+H^{+}+I^{-}\rightarrow I_2+H_2O$$

It is uncommon for reactions to take place involving trimolecular steps. The oxidation of nitrogen (II) oxide to nitrogen (IV) oxide,

$$2NO+O_2\rightarrow 2NO_2$$

is often classified as a trimolecular reaction but many believe it to involve two bimolecular reactions.

Triple Point. In a one component system three phases can only co-exist in equilibrium at a certain temperature and pressure. Thus, *e.g.*, in the system ice, water and water vapour, the three phases are in equilibrium at a pressure of 4 torr and a temperature of 0.0075°C. This point is termed the triple point for water.

Triplet. Three closely spaced transitions in a spectrum.

Tunnel Effect. An effect in which electrons are able to tunnel through a narrow potential barrier that would constitute a forbidden region if the electrons were treated as classical particles. That there is a finite probability of an electron tunnelling from one classically allowed region to another arises as a consequence of quantum mechanics. The effect is made use of in the tunnel diode.

Twaddell Hydrometers. Hydrometers graduated in degrees Twaddell, an arbitrary scale. They are still used, occasionally, for liquids of density greater than 1. Degrees Twaddell and relative density are related by the equation

$$\text{relative density} = \frac{200 + n^{\circ}\text{Tw}}{200}$$

Twinning. A term used in crystallography to denote a set of crystals that have one or more planes or faces in common but whose main developments away from that plane are not those of a single crystal.

Two-Component System. A system consisting of two components ; most commonly met in problems involving the distillation of two liquids.

Tyndall Effect. When a beam of light is passed through a disperse system, the path of light is visible because of the scattering caused by the particles of the disperse phase. This is the Tyndall effect, which is shown both by colloidal sols as well as coarse suspensions. A common example is the illumination of dust particles by sunlight. The light produced by Tyndall scattering is polarized. The effect is greatest for light of short wavelengths and the Tyndall cone is thus bluish in colour.

U

Ultracentrifuge. A device used to bring about the sedimentation of colloidal sols by subjecting them to forces many times that of gravity. An ultracentrifuge may be used to determine the molecular weights and size distribution of sol particles.

Ultrafiltration. A method of filtration in which particles of colloidal dimensions are separated from molecular and ionic substances by drawing the sol liquid through a membrane whose capillaries are very small. Membranes of collodion supported on filter paper are often used and these may be formed with different porosities so that particles of different maximum sizes may be retained. The process may also be used to cause the sedimentation of large organic molecules in aqueous solution and of heavy inorganic ions such as caesium.

The mechanism of ultrafiltration is not simply a sieve effect, but depends also upon the electrical conditions of both the membrane and the colloid.

Ultramicroscope. A form of microscope that uses the Tyndall effect to reveal the presence of particles that cannot be seen with a normal optical microscope. Colloidal particles, smoke particles, etc., are suspended in a liquid or gas in a cell with a black background and illuminated by an intense cone of light that enters the cell from the side and has its apex in the field of view. The particles then produce diffraction-ring systems, appearing as bright specks on the dark background.

Ultra-violet Light, U.V. Radiation which occurs beyond the visible violet light in the spectrum, *i.e.*, light of wavelength less than about 3600 A, is termed ultra-violet light. Radiation of yet shorter wavelength is called X-rays. Ultra-violet light possesses much greater energy than visible radiation, and is generally much more effective in inducing photochemical reactions, but possesses much less penetrating power. Ordinary glass is opaque to light of wavelength less than about 3600 A°. Quartz is transparent down to about 1800 A°, and is therefore used for making prisms and lenses in ultra-violet optical apparatus.

Uniaxial. Applied to crystals which have one principal axis, the other two axes being equivalent. Hexagonal, tetragonal and rhombohedral crystals are uniaxial.

Unimolecular. Describing a reaction (or step) in which only one molecule is involved. For example, radioactive decay is a unimolecular reaction :

$$Ra \rightarrow Rn + \alpha$$

Only one atom is involved in each disintegration.

In a unimolecular chemical reaction, the molecule acquires the necessary energy to become activated and then decomposes. The majority of reactions involve only uni- or bimolecular steps. The following reaations are all unimolecular :

$$N_2O_4 \rightarrow 2NO_2$$
$$PCl_5 \rightarrow PCl_3+Cl_2$$
$$CH_3CH_2Cl \rightarrow C_2H_4+HCl$$

Unimolecular Films. When certain insoluble oils and fats (*e.g.*, stearic acid) are dispersed on the surface of water it is possible to prepare these as unimolecular monolayers, as shown by the classic studies of Langmuir, N.K. Adam and others. In these unimolecular films the hydrophobic group is oriented away from the water ; the hydrophilic group is orientated towards it. In some cases the molecules form close-packed arrays, whilst in other cases the molecules are well-separated, thus forming different types of two-dimensional solids, liquids or gases.

Unimolecular Reaction. A chemical reaction or step involving only one molecule. An example is the decomposition of dinitrogen tetroxide :

$$N_2O_4 \rightarrow 2NO_2$$

Molecules colliding with other molecules acquire sufficient activation energy to react, and the activated complex only involves the atoms of a single molecule.

Unit. A reference value of a quantity used to express other values of the same quantity. See *also* SI units.

Unit Cell. The smallest group of atoms, ions, or molecules that when repeated at regular intervals in three dimensions, will produce the lattice of a crystal system. There are seven basic types of unit cells, which result in the seven crystal systems.

Unstable Equilibrium. See equilibrium.

V

Vacancy. See defect.

Vacuum. A space having gas below atmospheric pressure. A perfect vacuum contains no matter at all, but for practical purposes *soft* (*low*) *vacuum* is usually defined as down to about 10^{-2} pascal, and *hard* (*high*) *vacuum* as below this. *Ultrahigh vacuum* is lower than 10^{-7} pascal.

Vacuum Distillation. Distillation under reduced pressure. The depression in the boiling point of the substance distilled implies that the temperature is lower, which may prevent the substance from decomposing.

Vacuum Pump. A pump which is used to reduce the gas pressure in a container. The normal laboratory rotary oil-seal pump can maintain a pressure of 10^{-1} Pa. For pressures down to 10^{-7} Pa a diffusion pump is required. Ion pumps can achieve a pressure of 10^{-9} Pa and a cryogenic pump combined with a diffusion pump can reach 10^{-13} Pa.

Van Der Graaff Generator. An electrostatic machine used for accelerating positive ions.

Van Der Waals' Adsorption. An alternative name for the physical adsorption of species upon an adsorbent in which the adsorbate-adsorbent bonds are van der Waals' forces. See adsorption, physical.

Van Der Waals' Equation. Because the molecules in a gas have a definite volume and exert forces of attraction on each other, real gases do not obey the ideal gas laws to a high degree of accuracy. The effects of real gases were taken into account by van der Waals in his equation of state :

$$\left(P+n^2\frac{a}{V^2}\right)(V-nb)=nRT$$

where *a* and *b* are constants characteristic of the gas under consideration. The term (a/v^2) allows for the reduction of pressure from the ideal value due to intermolecular attractive forces ; *b* is the 'co-volume' and allows for the volume occupied by the molecules, it has a value of about four times the actual volume of the molecules.

Van Der Waals' Force. An attractive force between atoms or molecules, named after J D. van der Waals (1837-1923). The force accounts for the term a/V^2 in the van der Waals equation see equation of state). These forces are much weaker than those arising from valence bonds and are inversely proportional to the seventh power of the distance between the atoms or molecules. They are the forces responsible for non-idealbe haviour of gases and for the lattice energy of molecular crystals. There are three factors cavsing such forces : (1) dipole-dipole interaction, *i.e.*, electrostatic attractions between two molecules with permanent dipole moments ; (2) dipole-induced dipole interactions, in which the dipole of one molecule polarizes a neighbouring molecule ; (3) dispersion forces arising because of small instantaneous dipoles in atoms.

Van't Hoff Factor. Symbol : *i*. The ratio of the number of particles present in a solution to the number of undissociated molecules added. It is used in studies of colligative properties, which depend on the number of entities present. For example, if *n* moles of a compound are dissolved and dissociation into ions occurs, then the number of particles present will be *in*. Osmotic pressure (II), for instance will be given by the equation

Van't Hoff Isochore. The equation :

$$(d\log_e K)/dT = \triangle H/RT^2$$

showing how the equilibrium constant K, of a reaction varies with thermodynamic temperature, T. $\triangle$H is the enthalpy of reaction.

Van't Hoff Isochore. The name given to the relation

$$\frac{d \ln K}{dT} = \frac{\triangle H}{RT^2}$$

where K is the equilibrium constant of a reversible reaction, $\triangle$H the enthalpy of reaction, T the absolute temperature and

R the gas constant. This equation permits the calculation of enthalpies of reaction by measuring the variation of K with temperature.

Van't Hoff Isotherm. In a reversible chemical reaction of the type

$$aA + bB \rightleftharpoons cC + dD$$

the free energy change ΔG is given by the equation

$$\Delta G = -RT \ln K + RT \ln \left[\frac{[C]^c [D]^d}{[A]^a [B]^b} \right]$$

where R is the gas constant, T the absolute temperature, K the equilibrium constant, [A], [B], etc., the activities of A, B, etc., under the reaction conditions.

Vaporization. The process by which a liquid or solid is converted into a gas or vapour by heat. Unlike boiling, which occurs at a fixed temperature, vaporization can occur at any temperature. Its rate increases as the temperature rises.

Vapour. A gas formed by the vaporization of a solid or liquid. Some particles near the surface of a liquid acquire sufficient energy in collisions with other particles to escape from the liquid and enter the vapour ; some particles in the vapour lose energy in collisions and re-enter the liquid. At a given temperature an equilibrium is established. which determines the vapour pressure of the liquid at that temperature.

Vapour Density. For a given substance the ratio of the weight of a given volume of its vapour to that of the same volume of its vapour to that of the same volume of hydrogen, measured under identical conditions of temperature and pressure, is called the vapour density of the substance.

Vapour Pressure. Over every liquid or solid there is a certain pressure of its vapour. In a closed vessel after sufficient time, equilibrium is established, so that as many molecules leave the liquid surface to form vapour as return from the vapour to the liquid. At equilibrium at the specified temperature the pressure exerted by the liquid or solid is termed the vapour pressure at that particular temperature.

Vegard's Law. The statement that in phases in which there is a range of composition of solid solution then the cell dimensions vary linearly with the molar proportions of the constituents. The law rarely followed exactly.

Velocity of Reaction. The velocity of a chemical reaction is the amount of reactants transformed, or of product produced, per unit time. Thus velocity is usually expressed in gram moles, or gram atoms, per second.

Vibrational Spectrum. When a molecule absorbs energy the vibrational energy of the constituent atoms relative to each other may be increased. Conversely, when a transition from a higher vibrational level to a lower one occurs, it corresponds to the emission of energy, generally as radiation, of frequency v, where the energy loss E is related to v by the equation :

$$E = hv$$

Vibrational transitions, usually associated with simultaneous rotational transitions, occur in the i.r. and near i.r., and give rise to bands characteristic of the vibrational and rotational changes. These bands constitute the vibrational-rotational spectrum.

Vibrational-Rotational Spectrum. See vibrational spectrum.

Virial Equation. For an ideal gas the compressibility factor, PV/RT, is unity. However, for real gases the compressibility factor is a function of temperature and volume which may be expressed in terms of a virial equation of the form

$$\frac{PV}{RT} = 1 + \frac{B}{V} + \frac{C}{V^2} + \ldots$$

where B and C are functions of temperature, referred to as the second and third virial coefficients respectively.

Viscosity. All fluids show a resistance to flow, called viscosity. Mobile liquids, *e.g.*, water, have a low viscosity whereas liquids such as oil or treacle have a high viscosity and flow only with difficulty. Viscosity measurements are often used to determine the molecular weights of polymeric substances. Viscosities of different liquids are often compared in terms of viscosity coefficients.

For a Newtonian fluid, the force, F needed to maintain a velocity gradient, dv/dx, between adjacent planes of a fluid of area A is given by : $F=\eta A(dv/dx)$, where η is a constant, called the coefficient of viscosity. In SI units it has the unit pascal second (in the c.g.s. system it was measured in poise). Non-Newtonian fiuids, such as clays, do not conform to this simple model. See also kinematic viscosity.

Viscosity, Coefficient of. The force required per unit area of fluid to maintain unit difference of velocity between two parallel planes in the fluid at a fixed distance apart. For plates one centimetre apart, the unity of viscosity is the poise.

Viscosity Index (VI). An empirical number expressing the temperature-viscosity relationship of an oil. A low viscosity index is an indication of a large change of viscosity with temperature and vice versa.

Viscosity index is determined by measuring the viscosity of an oil at two temperatures and comparing the figures with those for standard oils of VI=100 and VI=0.

Volatile. Easily converted into a vapour.

Volt. Symbol : V. The SI unit of electrical potential, potential difference, and e.m.f., defined as the potential difference between two points in a circuit between which a constant current of one ampere flows when the power dissipated is one watt. $1\ V=1\ J\ C^{-1}$.

Voltaic Cell (Galvanic Cell). A device that produces an e.m.f. as a result of chemical reactions that take place within it. These reactions occur at the surfaces of two electrodes, each of which dips into an electrolyte. The first voltaic cell, devised by Alessandro Volta (1745-1827), had electrodes of two different metals dipping into brine. See primary cell ; secondary cell.

Voltameter (Coulometer). 1. An electrolytic cell formerly used to measure quantity of electric charge. The increase in mass (m) of the cathode of the cell as a result of the deposition on it of a metal from a solution of its salt enables the charge (Q) to be determined from the relationship $Q=m/z$, where z is the electrochemical equivalent of the metal. 2. Any other type of electrolytic cell used for measurement.

W

Watt Symbol : W. The SI unit of power, defined as a power of one joule per second. In electrical contexts it is equal to the rate of energy transformation by an electric current of one ampere flowing through a conductor the ends of which are maintained at a potential difference of one volt. The unit is named after James Watt (1736-1819).

Wavelength, λ. The distance between corresponding points in the profile of wave motion.

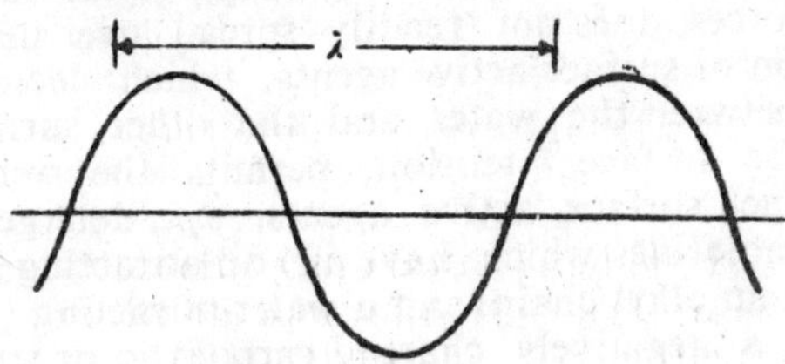

Wave Mechanics. See quantum theory.

Weak Acid. An acid that is only partially dissociated in aqueous solution.

Weber Symbol : Wb. The SI unit of magnetic flux, equal to the magnetic flux that, linking a circuit of one turn, produces an e.m.f. of one volt when reduced to zero at a uniform rate in one second. 1 Wb=1 V s.

Weston Cell (Cadmium Cell. A type of primary voltaic cell, which is used as a standard ; it produces a constant e.m.f. of 1.0186 volts at 20°C. The cell is usually made in an H-shaped glass vessel with a mercury anode covered with a paste of cadmium sulphate and mercury(I) sulphate in one leg and a cadimum amalgam cathode covered with cadmium sulphate in the other leg. The electrolyte, which connects the two electrodes by means of the bar of the H, is a saturated solution of cadmium sulphate. In some cells sulphuric acid is added to prevent the hydrolysis of mercury sulphate.

Weston Cadmium Cell. A standard cell that producess a constant e.m.f. of 1.0186 volts at 20°C. It consisting of an H-shaped glass vessel containing a negative cadmium-mercury amalgum electrode in one leg and a positive mercury electrode in the other. The electrolyte—saturated cadmium sulphate solution—fills the horizontal bar of the vessel to connect the two electrodes. The e.m.f. of the cell varies very little with temperature, being given by the equation $E=1.0186-0.000037\,(T-293)$, where T is the thermodynamic temperature.

Wet Cell. A type of cell, such as a car battery, in which the electrolyte is a liquid solution. See *also* Leclanche cell.

Wetting Agents. Water, because of its powerful intermolecular attractive forces, does not readily spread over many surfaces. The addition of surface active agents, which decrease the contact angle between the water and the other surface, thereby lowering the surface tension, permits the wetting of the surface. Such surface active agents, *e.g.*, detergents, usually consist of molecules which have an oil-attracting hydrophobic group (*e.g*, an alkyl chain) and a water-attracting (hydrophilic) group (*e.g.*, a negatively charged carboxylic or sulphonic acid group).

Work Function. A quantity that determines the extent to which thermionic or photoelectric emission will occur according to the Richardson equation or Einstein's photoelectric equation. It is sometimes expressed as a potential difference (symbol ϕ) in volts and sometimes as the work to be done by the exciting electron (symbol W) in electronvolts or joules. The former has been called the *work function potential* and the latter the *work function energy.*

X

Xerogels. A method of classification of gels. Xerogels are relatively free of the dispersion medium ; lyogels are rich in the dispersant.

Xerography. A reprographic process in which a photoconductor (generally Se) on a surface is charged, exposed to the image, which discharges all except the image, removal of Se from the non-image areas, developing the image with an oppositely charged pigment, transferring the image to paper electrostatically and fixing the image on paper by baking.

X-ray. Electromagnetic radiation of wave length *c.* 1A°. X-rays are generated in various ways, including the bombarding of solids with electrons, when they are emitted as a result of electron transitions in the inner orbits of the atoms bombarded. Each element has a characteristic X-ray spectrum.

X-rays may be detected either photographically or with an ionization counter. They have great penetrating power which increases with their frequency, and owing to this are used to photograph the interior of many solid objects, stably the human body and in monitoring for faults in construction.

X-rays find wide applications in X-ray photography and in crystallography. Prolonged exposure of the human body to the rays induces a dangerous form of dermatitis, and even sterility, but controlled exposures are applied to alleviate cancer.

X-ray Crystallography. The use of X-ray diffraction to determine the structure of crystals or molecules. The technique involves directing a beam of X-rays at a crystalline sample and recording the diffracted X-rays on a photographic plate. The diffraction pattern consists of a pattern of spots on the plate, and the crystal structure can be worked out from the positions and intensities of the diffraction spots. X-rays are diffracted by the electrons in the molecules and if molecular crystals of a compound are used, the electron density distribution in the molecule can be determined.

X-ray Diffraction. A powerful technique for the study of crystal structure, discovered by von Laue (1912) and developed for crystal analysis by W.H. and W.L. Bragg (1912-1913). The atomic nuclei in a crystal lattice act as diffraction gratings; the rows of atoms have spacings of a few Angstrom units, which are comparable with the wavelength of X-rays. Strong scattering of the rays by the crystal therefore occurs in certain directions, according to Bragg's equation. Various techniques are available for applying X-ray diffraction to the study of single crystal powders, fibres, etc., and by computation it is possible to work out 3-dimensional electron-density maps of the solid lattice from the recorded X-ray patterns.

X-ray Fluorescence. The emission of X-rays from excited atoms produced by the impact of high-energy electrons, other particles, or a primary beam of other X-rays The wavelengths of the fluorescent X-rays can be measured by an X-ray spectrometer as a means of chemical analysis. X-ray fluorescence is used in such techniques as electron-probe mieroanalysis.

X-ray Spectrometer. An apparatus used in the X-ray study of crystals in which a fine beam of monochromatic X-rays impings at a measured angle on the face of a crystal mounted in its path, and in which the intensity of the X-rays diffracted in various directions by the crystal is measured with an ionization chamber mounted on an arm of the spectrometer table, or is recorded photographically.

X-ray Spectroscopy. Analytical method by which a sample is irradiated with X-rays, characteristic radiation being emitted after scattering from the specimen. The detection limits for various elements are of the ordering cm^{-2}.

X-ray Tube. The apparatus employed for producing X-rays. It consists essentially of an aluminium electrode, the cathode and an anode (always called the anticathode in X-ray work), the surface of which is arranged at an angle to the axis of the tube. Under the influence of a high potential, cathode rays emitted by the cathode strike the anticathode, which emits X-rays. These are directed out of the tube by the sloping anticathode.

Z

Zero Order. Describing a chemical reaction in which the rate of reaction is independent of the concentration of a reactant : *i.e.*,

$$rate = k[X]^0$$

The concentration of the reactant remains constant for a period of time although other reactants are being consumed. The hydrolysis of 2-bromo-2-methyl propane using aqueous alkali has a rate expression.

$$rate = k[\text{2-bromo-2-methyl propane}]$$

i.e. the reaction is zero order with respect to the concentration of the alkali. The rate constant for a zero reaction has the units mol dm^{-3} s^{-1}.

Zeroth Law of Thermodynamics. See thermodynamics.

Zero Point Energy. The energy possessed by the atoms and molecules of a substance at absolute zero (0 K).

Zwitterion. (ampholyte ion). An ion that has both a positive and negative charge on the same species. Zwitterions occur when a molecule contains both a basic group and an acidic group ; formation of the ion can be regarded as an internal acid-base reaction. For example, aminoethanoic acid (glycne) has the formula $H_2N.CH_2.COOH$. Under neutral conditions it exists as the zwitterion $^+H_3N.CH_2.COO^-$, which can be formed by transfer of a porton from the carboxyl group to the amine group. At low pH (acidic conditions) the ion $^+H_3N.CH_2.COOH$ is formed ; at high pH (basic conditions) $H_2N.CH_2.COO^-$ is formed. See *also* amino acid.